EasieR
Practical Applications of Machine Learning Algorithms in R

Darrin Thomas

EasieR
Practical Applications of Machine Learning Algorithms in R

Darrin Thomas
Asia-Pacific International University

SuJinSoLa
Saraburi, Thailand

Layout: Darrin Thomas
Photo Researcher: Darrin Thomas
Cover Design: Darrin Thomas

Copyright © 2017 ERT Group, publishing as SuJinSoLa, Saraburi, Thailand. All rights reserved.

SuJinSoLa

ISBN-13: 978-1976556814
ISBN-10: 1976556813

A   the Author

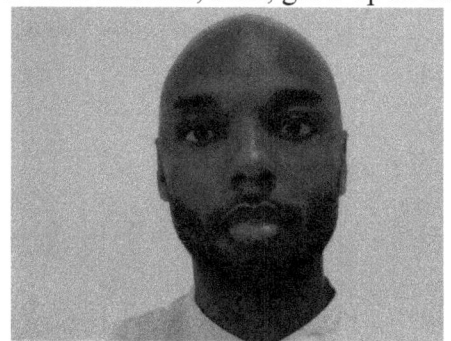

Darrin Thomas, PhD, grew up in Sacramento, California and has over ten years of experience as a teacher and lecturer from Kindergarten to graduate school. He completed his bachelor and master degree in saxophone performance at California State University Sacramento. After working as a substitute teacher, he completed a credential in teaching at Pacific Union College. He then worked as a music teacher before moving to Thailand to work as a lecturer in Education/Psychology Department at Asia-Pacific International University (APIU). While overseas, Dr. Darrin completed his master degree in education at APIU. He then moved to the Philippines and completed his doctoral degree in education at Adventist International Institute of Advanced Studies. Currently, Dr. Darrin is a lecturer at Asia-Pacific International University. His enthusiasm for machine learning has led to works involving many different algorithms applied in an educational context.

DEDICATION

To my wife and children

Table of Contents

About the Author iii
Preface ix

Chapter 1 Decision Trees 1
Chapter 2 Numeric Decision Trees 9
Chapter 3 Random Forest 19
Chapter 4 Classification Rules 26
Chapter 5 Support Vector Machines 36
Chapter 6 Support Vector Machines Numeric Prediction 43
Chapter 7 Artificial Neural Networks 52
Chapter 8 Artificial Neural Networks Numeric Prediction 60
Chapter 9 H2O Deep Learning 67
Chapter 10 Evaluating Model Performance 80
Chapter 11 Improving Model Performance 86

Detail Table of Contents

Chapter 1 Decision Trees 1
    Chapter Objectives 1
    Definition of Decision Trees 1
    How Decision Trees Work 1
    Types of Decision Trees 2
    Pros and Cons of Decision Trees 3
    Data Preparation 3
    Model Development 5
    Model Testing 7
    Conclusion 7

Chapter 2 Numeric Decision Trees 9
    Chapter Objectives 9
    Types of Numeric Prediction Trees 9
    Pros and Cons of Numeric Prediction Trees 10
    Data Preparation 10
    Model Development 13
    Model Testing 14
    Modal Tree Development and Testing 16
    Conclusion 18

Chapter 3 Random Forest 19
    Chapter Objectives 19
    Concerns with Decision Trees 19
    Random Forest 19
    Pros and Cons of Random Forest 20
    Data Preparation 20
    Random Forest Classification 21
        Model Development 21
        Model Testing 22
    Random Forest Regression 23
        Model Development 23
        Model Testing 24
    Conclusion 26

Chapter 4 Classification Rules 27
    Chapter Objectives 27
    Classification Rules 27
    The Algorithm 28
    Pros and Cons of Classification Rules 28
    Data Preparation 28
    One Rule Algorithm 32
        Model Development 32
        Model Testing 33
    JRip Algorithm 33
        Model Development 33
        Model Testing 34
    Conclusion 35

Chapter 5 Support Vector Machines 36

- Chapter Objectives ... 36
- Defining Support Vector Machines ... 36
- Pros and Cons of Support Vector Machines ... 37
- Data Preparation ... 37
  - Linear Kernel ... 41
    - Model Development ... 41
    - Model Testing ... 41
  - Radial Kernel ... 42
    - Model Development and Testing ... 42
  - Conclusion ... 42
- **Chapter 6 Support Vector Machine Numeric Prediction** ... 43
  - Chapter Objectives ... 43
  - More on SVM ... 43
  - Data Preparation ... 44
  - Model Development and Testing with Kernels ... 45
    - Linear Kernel ... 46
    - Polynomial Kernel ... 48
    - Radial Kernel ... 49
    - Sigmoid Kernel ... 50
  - Conclusion ... 51
- **Chapter 7 Artificial Neural Networks** ... 52
  - Chapter Objectives ... 52
  - The Human Mind and the Artificial One ... 52
  - Activation Function ... 53
  - ANN Design ... 53
  - Pros and Cons of ANNs ... 54
  - Data Preparation ... 54
  - Single Hidden Layer ... 55
    - Model Development ... 55
    - Model Testing ... 56
  - Multiple Hidden Layers ... 57
    - Model Development ... 57
    - Model Testing ... 58
  - Conclusion ... 59
- **Chapter 8 Artificial Neural Networks Numeric Prediction** ... 60
  - Chapter Objectives ... 60
  - Data Preparation ... 60
  - Single Node ... 63
    - Model Development ... 63
    - Model Testing ... 64
  - Multiple Nodes ... 65
    - Model Development ... 65
    - Model Testing ... 65
  - Conclusion ... 66
- **Chapter 9 H20 Deep Learning** ... 67
  - Chapter Objectives ... 67
  - Data Preparation ... 68
  - Classification Model ... 72

  Model Development and Testing 72
 Numeric Prediction 75
  Model Development and Testing 75
 Conclusion 79
Chapter 10 Evaluating Model Performance 80
 Chapter Objectives 80
 Confusion Matrix 80
 Assessing Models with Confusion Matrix Outputs 83
 Numeric Prediction 85
 Conclusion 85
Chapter 11 Improving Model Performance 86
 Chapter Objectives 86
 Parameters 86
 Automatically Tuned Model 87
 Custom Tuned Model 88
 Conclusion 90

Preface

The purpose of this text is to provide a practical explanation of the use of R for machine learning purposes. There is no detailed explanation of the foundational theoretical principles of algorithms. This is a book for practitioners who want to get things done rather than dwell on the "why" of theories. Theories are important but not necessarily for everyone.

My interest in machine learning has grown as I have seen the discipline mature. I came into this field from as an indirect result of my graduate studies. Like many other students, I was looking for an alternative to purchasing SPSS for personal use. This led to R and developing an understanding for how this open-source resource works.

The focus within this text is primarily on commonly used algorithms. Examples include decision trees, support vector machines, and artificial neural networks. Each chapter provides examples of how to run these algorithms on a dataset in R.

There are several assumptions that this text makes. One, the reader is already familiar with the process of developing models in a machine-learning context. You know how data needs to be prepared, how a model is developed, the need for evaluating the model with the test set, and how to adjust parameters if necessary.

Another assumption of this text is that the readers are already familiar with R and R studio. What this means is that you are able to read and interpret the code and have experience with the mathematics and computer coding skills required for machine learning. I make no claim to being an expert in R. As I learn, I learn more and more about what I do not know. However, I do know how to get things done when using this programming language.

Chapter One: Decision Trees

Decision trees provide the user with a visual of how to make a choice based on certain criteria. The criteria for what choice to make is determined mathematically. A great deal of analysis is done with the use of decision trees in many different problems in research with a plethora of applications. The goal is to sub-divide the sample into subsets that are highly within group similar.

Chapter Objectives
The objectives of this chapter are as follows.
- Define what decision trees are
- Explain the types of decision trees
- Physical traits of decision trees
- How decision trees work
- Pros and cons of decision trees
- Physical Traits of a Decision Tree

Definition of Decision Trees
Decision trees consist of what is called a tree structure. The tree structure consists of a root node, decision nodes, branches, and leaf nodes. A root node is the initial decision made in the tree. This depends on which feature the algorithm selects first. Following the root node, the tree splits into various branches. Each branch leads to an additional decision node where the data is further subdivided. At the bottom of a tree are the terminal node(s) that are also called leaf nodes. Decision trees look a lot like an organizational chart. Figure 1.1 is a visual of the components of a decision tree.

How Decision Trees Work
There are two commonly used algorithms when developing decision trees. Decision trees using the "rpart" algorithm in R use a heuristic called recursive partitioning. What this does is it splits the overall data set into smaller and smaller subsets until each subset is as close to pure (having the same characteristics) as possible. This process is also known as divide and conquer.

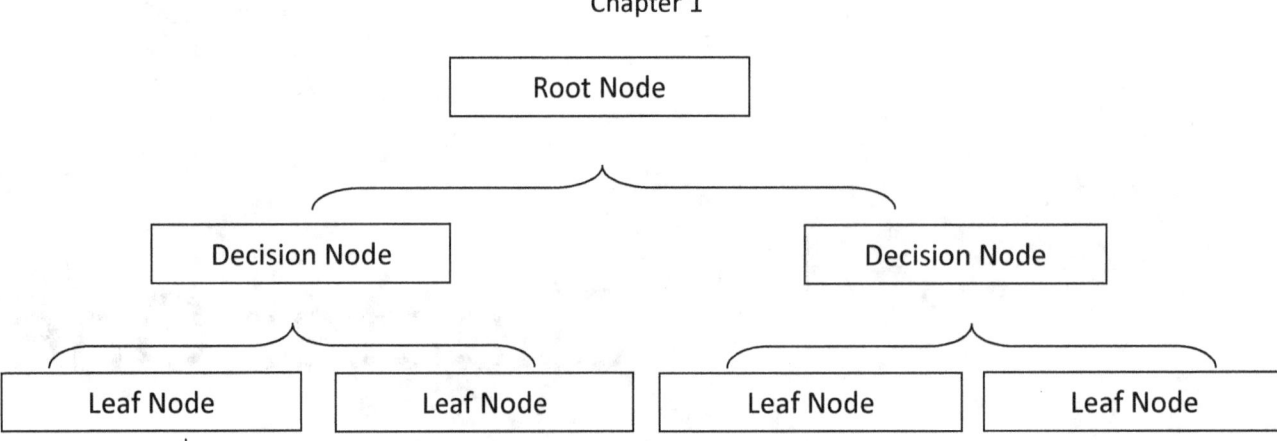

Figure 1.1 Example of Decision tree

Other popular decision tree algorithms include C5.0, Quest, and CTree. The differences between these algorithms is small in terms of performance and it is the context that determines which is most appropriate. The details of this are beyond the scope of this book but the main differences among the algorithms are how the decision nodes are selected. In my experience, the use of the "rpart" algorithm is easier for beginners because you can make highly clear and informative visual aids using other packages.

The mathematics for deciding how to split the data is based on a concept called entropy, which measures the purity of a potential decision node. The lower the entropy score the more pure the decision node is. The entropy can range from 0 (most pure) to 1 (most impure).

Types of Decision Trees
There are two main types of decision trees. These two types are classification and numeric trees. The ultimate goal of a classification tree is to classify an example categorically. For a numeric tree, the goal is to make a final numerical or interval/ratio level calculation/prediction. What makes a decision tree a classification or numeric tree is the type of dependent variable in the model. Categorical dependent variables lead to classification while interval/ratio dependent variable leads to numeric. Another term for numeric trees is regression trees because regression trees make numeric predictions as linear regression does.

The setup for the coding for either a classification or numeric tree is the same. However, the interpretation is different because different metrics are used to measure the quality of the model. Classification trees are often focused on accuracy while numeric trees are focused on reducing error. You can use a mixture of variables in both types. As I have already mentioned, remember that the dependent variable determines the tree type whether classification or numeric prediction.

A variation on these two types of decisions trees is random forest. Random forest involves the creation of many trees that are then used to "vote" for the correct value. This is an example of ensemble learning in that a group of trees decides on how to classify an example or to calculate a value. You will learn more about this in the chapter on random forest.

Pros and Cons of Decision Trees

The pros of decision trees include their versatile nature. Decision trees can deal with all types of data as well as missing data. Furthermore, this approach learns automatically and only uses the most important features. Lastly, a deep understanding of mathematics is not necessary to use this method in comparison to more complex models.

Some problems with decision trees are that the can easily overfit the data. This means that the tree does not generalize well to other datasets. In addition, a large complex tree can be hard to interpret, which may be yet another indication of overfitting.

Data Preparation

The rest of this chapter is dedicated to providing an example of a classification tree. We are going to use the 'Wage' dataset found in the "ISLR" package. Our goal is going to be to divide workers based on their educational level using age, jobclass, and wages as independent variables.

Before we go further, the example in this chapter is somewhat unusual because or dependent variable "education" has several levels to it. Often, the use of classification trees has a dependent variable with only two factors. This allows you to calculate various metrics involving the true/false positive/negatives. However, the purpose here was to do a realistic analysis rather than standard practice. Therefore, we can only look at accuracy as our criteria for model quality. Below is the code.

```
library(ISLR);library(caret);library(e1071);library(rattle);library(rpart);
library(rpart.plot)
data("Wage")
inTrain<-createDataPartition(y=Wage$education,p=0.7, list=FALSE)
trainingset <- Wage[inTrain, ]
testingset <- Wage[-inTrain, ]
```

We began by loading several needed libraries. Next, we loaded the data "Wage." After this, we used the "createDataPartition" function to split the data into a training and test set. The split was based on our dependent variable "education." 70% of the data was set aside for the training and 30% for testing. Lastly, we subsetted the "Wage" dataset to create our testing and training set.

We will now make a plot of the variables to see how they look. Below is the code followed by the plot. Please note that education is divided into 5 groups as indicated in the chart.

```
table(Wage$education)
## 
##         1. < HS Grad          2. HS Grad      3. Some College
##                  268                 971                  650
##      4. College Grad  5. Advanced Degree
##                  685                 426
table(Wage$jobclass)
## 
##   1. Industrial 2. Information
##            1544           1456
```

The tables that we created gives us information on our nominal variables. The proportions are reasonable for both "education" and "wage." We will now look at histograms of the interval/ratio level variables. Figure 1.2 is the "age" variable and figure 1.3 is the "wage" variable.

`hist`(Wage$age)

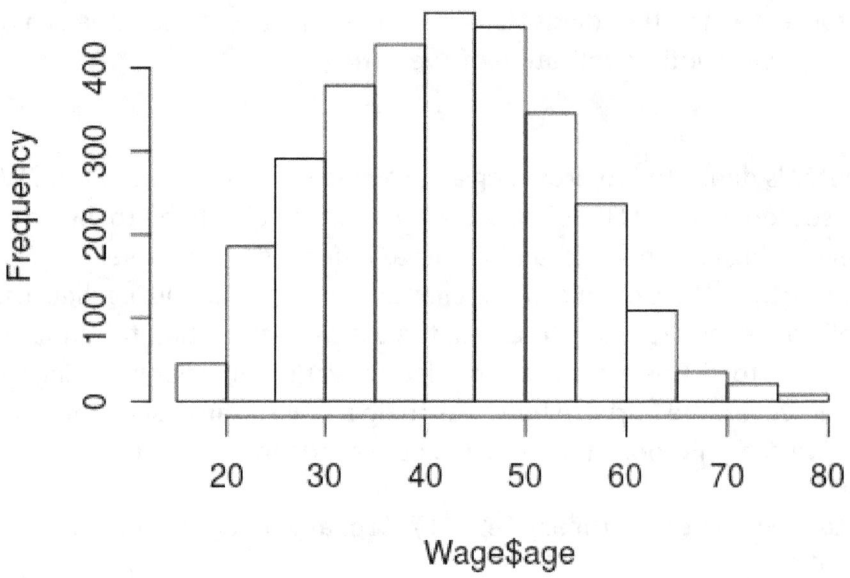

Figure 1.2 Age Histogram

`hist`(Wage$wage)

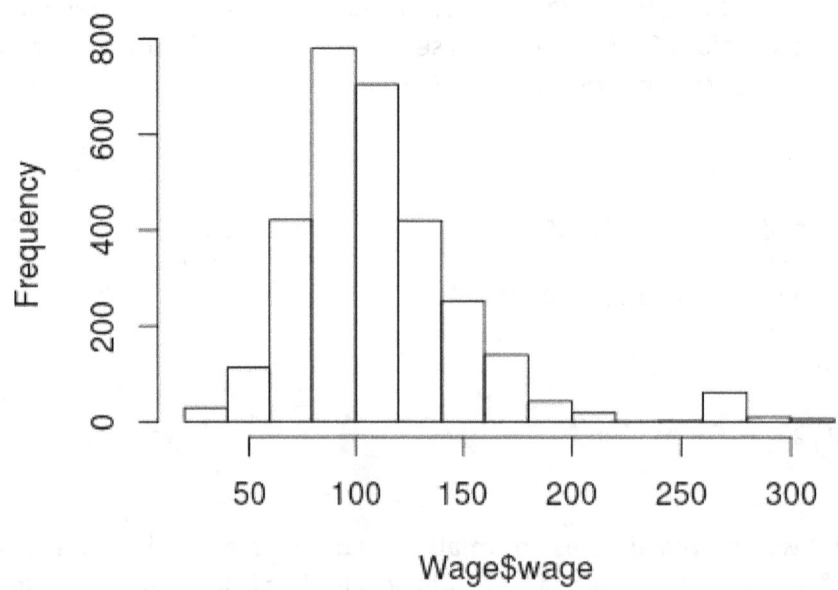

Figure 1.3 Wage Histogram

There doesn't appear to be any major problems with the data as everything looks mostly normal. However, decision trees can handle non-normal and messy data well. Therefore, we can continue to training the model.

Model Development

In the code below, we are using the "train" function from the "caret" package to create our classification tree. The code for the actual model is the same as that for making a regression model. We also need to indicate the method of analysis, which for us is "rpart." Lastly, we need to indicate what dataset to use.

```
set.seed(1)
TreeModel<-train(jobclass~age+wage+education, method='rpart', data=trainingset)
TreeModel
## CART
##
## 2102 samples
##    3 predictors
##    2 classes: '1. Industrial', '2. Information'
##
## No pre-processing
## Resampling: Bootstrapped (25 reps)
## Summary of sample sizes: 2102, 2102, 2102, 2102, 2102, 2102, ...
## Resampling results across tuning parameters:
##
##   cp           Accuracy   Kappa
##   0.007744434  0.6162249  0.2302851
##   0.102613746  0.5897821  0.1796914
##   0.148112294  0.5664217  0.1351289
##
## Accuracy was used to select the optimal model using  the largest value.
## The final value used for the model was cp = 0.007744434.
```

By default, the "caret" package will produce three different models. The difference in the models is based on how the tuning parameter is adjusted. For classification trees using the rpart algorithm, the tuning parameter is "cp". Cp stands for "complexity parameter" and is used to determine the growth of the tree.

Changing the value of "cp" affects the accuracy and kappa. Each model reports the accuracy of the model as well as the Kappa. The accuracy states how well the model predicted true positives/negatives. The kappa shares the same information but it calculates how well a model classifies while taking into account chance or luck. As such, the Kappa should be lower than the accuracy.

This process used by "caret" in developing multiple models for comparison purposes is an example of automatic tuning. The concept of automatic tuning will be addressed in detail later in chapter 11.

We now need to create a visual of the model. We will used the "fancyRpartPlot" from the "rattle" package. Figure 1.4 is the tree.

```
fancyRpartPlot(TreeModel$finalModel)
```

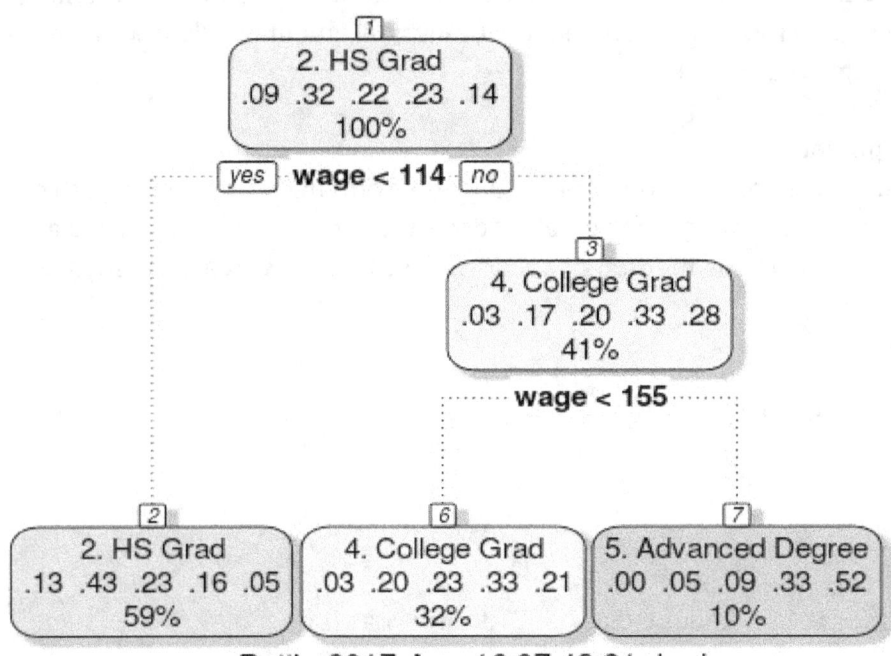

Figure 1.4 Classification tree

Here is what the chart means

At the top is node 1 or the "root node" which is called 'HS Grad" the decimals underneath is the percentage of the data that falls within the "HS Grad" category inside this node. For example, 9% of the examples in this node are people with less than a HS diploma who were misclassified as being a high school grad, 32% are HS grads who were correctly classified, 22% have some college (misclassified), 23% are college grads (misclassified), and 14% have an advanced degree (misclassified). The ordering of the percentages are based on the order of the factors in the variable. This first node is label "HS Grad" because HS grad is the largest proportion of examples in this node and before we apply any criteria for separating examples.

As the highest node, everything is classified as "HS grad" until we begin to apply our criteria. Underneath node 1 there is a leaf node (2. HS Grad) and a decision node (4. College Grad). To determine what node to take you must consider the wage. If a person makes less than 114 you go to the left if the make more than 114 they go to the right.

Node 2 indicates the percentage of the sample that was classified as HS grade regardless of education. Within this node, 13% of those with less than a HS diploma were misclassified as a HS Grade based on wage. 43% of those with a HS diploma were correctly classified as a HS grade based on income. You can see that there is some misclassification. For example, 16% of the examples in this node have a college degree but were classified as having only a high school diploma. This could be because despite their education they have a smaller salary than expected by this model. What we really want is 100% of all the examples in this node to be people with only a high school diploma. However, this is rarely attained and if it does happens probably indicates overfitting.

The percentage underneath the decimals in each node indicates the total amount of the sample placed in each node. For example, node 2, HS grad category has 59% of the entire sample within it. In other words, 59% of the total sample was classified as being high school graduates, 32% were classified as College grads and, 10% as having advance degrees. This adds up to about 100% but there is some rounding in the visual.

Another way to read a decision tree is the follow...
- If a person has a wage less than 114 then they are a high school graduate
- If a person has a wage greater than 114 but less than 155 then they are a college graduate
- If a person has a wage greater than 155 than they have an advance degree

Model Testing

You can predict individual values in the data set by using the 'predict' function with the test data as shown in the code below. Once the predictions are done, we use the "table" function to look at the classifications.

```
testmodel<-predict(TreeModel, newdata = testingset)
table(testmodel,testingset$education)
##
## testmodel          1. < HS Grad 2. HS Grad 3. Some College
##   1. < HS Grad                0         0              0
##   2. HS Grad                 71       228            111
##   3. Some College             0         0              0
##   4. College Grad             8        55             63
##   5. Advanced Degree          1         8             21
##
## testmodel          4. College Grad 5. Advanced Degree
##   1. < HS Grad                   0                  0
##   2. HS Grad                    76                 21
##   3. Some College                0                  0
##   4. College Grad               87                 57
##   5. Advanced Degree            42                 49
```

Obliviously there is a great deal of misclassifications. For example, 71 people who graduate from high school were classified as having less than a high school diploma. The numbers along the diagonal are correctly classified. For example, 228 examples were classified as high school graduates that were actually high school graduates. To calculate the accuracy, we need to take the total number correctly classified and divide it by the total number of examples. Below is the code

```
## 357/898
## [1] 0.3975501
```

Well, an accuracy of 39% is terrible. One of our biggest problems is that we did not correctly classify anybody with less than a high school diploma. This is a problem because there were 268 of these that were misclassified. Perhaps additional variables would have improved the model. However, the primary purpose of this exercise was to demonstrate how to conduct an analysis of data with classification trees.

Conclusion

Classification trees allow you to subdivide a dataset so that examples are classified in a way that they are as similar as possible. The example in this chapter provided a systematic approach using the "rpart" function. In the next chapter, we will continue looking at decision trees but this time we will do numeric prediction.

Chapter Two: Numeric Decision Trees

Decision trees are not limited to classification. We can also make numeric predictions with decision trees. In this chapter, we will look at two common types of numeric prediction trees and these are regression trees and modal trees.

Chapter Objectives
The objectives of this chapter are as follows.
- Explain the types of numeric prediction trees
- Explain the pros and cons of numeric prediction trees
- Prepare data for numeric prediction tree analysis
- Develop and test regression tree and modal tree models

Types of Numeric Prediction Trees
Regression trees have been around since the 1980's. They work by predicting the average value of specific examples that reach a given leaf in the tree. Despite their name, there is no regression involved with regression trees. Regression trees are straightforward to interpret but at the expense of accuracy due to the reliance on the mean as a tool for splitting the data.

Regression trees also use a concept called recursive partitioning (rpart). Recursive partitioning involves splitting features in a way that reduces the error the most. The splitting is also greedy which means that the algorithm will partition the data at one point without considering how it will affect future partitions. Ignoring how a current split affects the future splits can lead to unnecessary branches with high variance and low bias. We used this same algorithm for classifications trees in the previous chapter.

Modal trees are similar to regression trees but employs multiple regression with the examples at each leaf in a tree. This leads to many different regression models being used to split the data throughout a tree. This makes model trees hard to interpret and understand in comparison to regression trees. However, they are normally much more accurate than regression trees since they do not use the mean as the metric for splitting the data.

Both types of trees have the goal of making groups that are as homogeneous as possible. For classification trees, entropy is used to measure the homogeneity of groups. For numeric decision trees, the standard deviation reduction (SDR) is used. The SDR is simply a measure of how much the standard deviation is reduced due to a split in the data. The more the deviation is reduced the better as this indicates less variability in the data and more homogeneity.

Pros and Cons of Numeric Prediction Trees

Numeric prediction trees do not have the assumptions of linear regression. As such, they can be used to model non-normal and or non-linear data. In addition, if a dataset has a large number of feature variables, a numeric prediction tree can easily select the most appropriate ones automatically. Lastly, numeric prediction trees also do not need the model to be specified in advance of the analysis.

On the other hand, this form of analysis requires a large amount of data in the training set in order to develop a testable model. It is also hard to tell which variables are most important in shaping the outcome. Predictive performance can also be hurt when a particular example is assigned the mean of a node. This forced assignment is a loss of data such as turning continuous variables into categorical variables. Lastly, numeric prediction trees are sometimes hard to interpret. This naturally limits their usefulness among people who lack statistical training.

Data Preparation

In our example, we will try to predict how many kids a person has based on several independent variables in the "PSID" data set in the "Ecdat" package. Let's begin by loading the necessary packages and data set. The code is below

```
library(Ecdat);library(rpart);library(rpart.plot); library(RWeka)
data(PSID)
str(PSID)
## 'data.frame':    4856 obs. of  8 variables:
##  $ intnum  : int  4 4 4 4 5 6 6 7 7 7 ...
##  $ persnum : int  4 6 7 173 2 4 172 4 170 171 ...
##  $ age     : int  39 35 33 39 47 44 38 38 39 37 ...
##  $ educatn : int  12 12 12 10 9 12 16 9 12 11 ...
##  $ earnings: int  77250 12000 8000 15000 6500 6500 7000 5000 21000 0 ...
##  $ hours   : int  2940 2040 693 1904 1683 2024 1144 2080 2575 0 ...
##  $ kids    : int  2 2 1 2 5 2 3 4 3 5 ...
##  $ married : Factor w/ 7 levels "married","never married",..: 1 4 1 1 1 1 1 4 1 1 ...
summary(PSID)
##      intnum          persnum           age            educatn
##  Min.   :   4    Min.   :  1.00   Min.   :30.00   Min.   : 0.00
##  1st Qu.:1905    1st Qu.:  2.00   1st Qu.:34.00   1st Qu.:12.00
##  Median :5464    Median :  4.00   Median :38.00   Median :12.00
##  Mean   :4598    Mean   : 59.21   Mean   :38.46   Mean   :16.38
##  3rd Qu.:6655    3rd Qu.:170.00   3rd Qu.:43.00   3rd Qu.:14.00
##  Max.   :9306    Max.   :205.00   Max.   :50.00   Max.   :99.00
##                                                   NA's   :1
##     earnings          hours           kids              married
##  Min.   :    0    Min.   :   0    Min.   :0.000   married      :3071
##  1st Qu.:   85    1st Qu.:  32    1st Qu.:1.000   never married: 681
##  Median :11000    Median :1517   Median :2.000   widowed      :  90
```

```
##    Mean   : 14245     Mean   :1235     Mean   : 4.481     divorced    : 645
##    3rd Qu.: 22000     3rd Qu.:2000     3rd Qu.: 3.000     separated   : 317
##    Max.   :240000     Max.   :5160     Max.   :99.000     NA/DF       :   9
##                                                           no histories:  43
```

The variables "intnum" and "persnum" are for identification and are useless for our analysis. We will now explore our dataset by making histograms with the following code. Figure 2.1 is the histogram of the of our continuous variables.

```
hist(PSID$age)
hist(PSID$earnings)
hist(PSID$hours)
hist(PSID$kids)
hist(PSID$educatn)
table(PSID$married)
##
##       married  never married      widowed       divorced     separated
##          3071             681           90            645           317
##         NA/DF   no histories
##             9             43
```

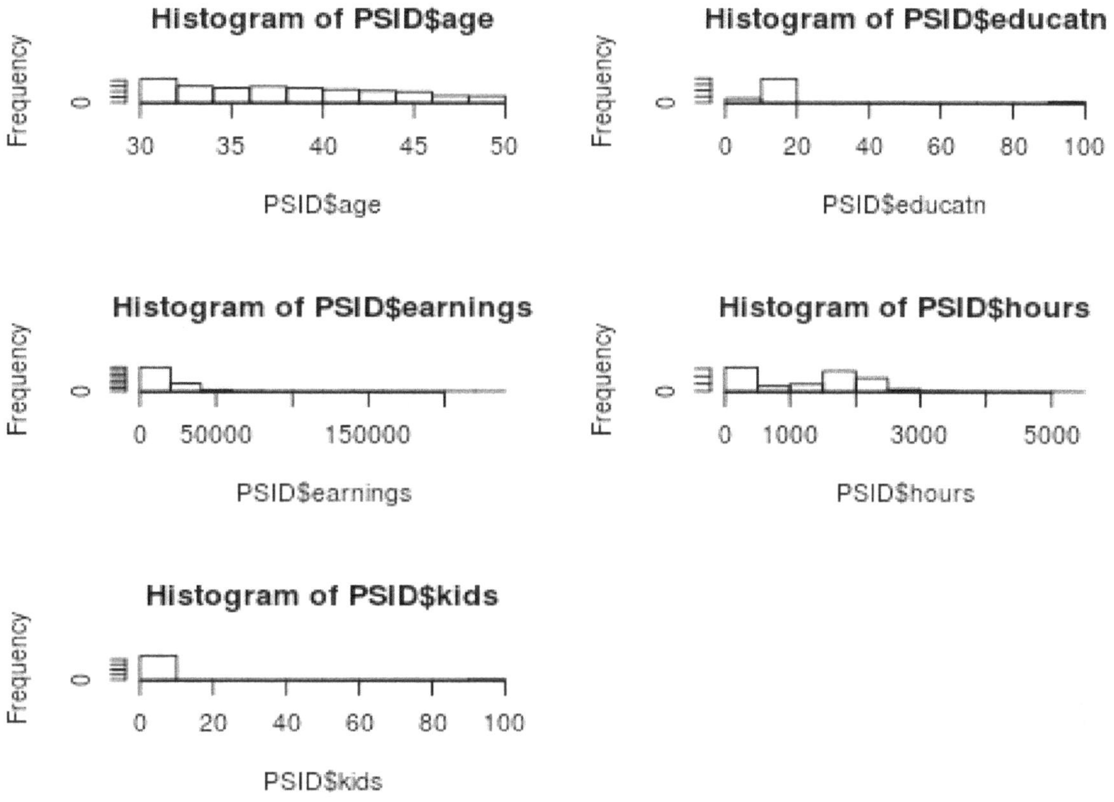

Fig 2.1: Continuous variables histogram

Almost all of the variables are non-normal. However, this is not a problem when using regression trees. There are some major problems with the "kids" and "educatn" variables. Each of these variables has values at 98 and 99. When the data for this survey was collected 98 meant the respondent did not know the answer and a 99 meant they did not want to say. Since both of these variables are numerical we have to do something with them so they do not ruin our analysis. As we know, it would be unusual for someone to have 98 or 99 children or have completed grade 98 or 99.

We are going to recode all values equal to or greater than 98 as 3 for the "kids" variable. The number 3 means they have 3 kids. This number was picked because it was the most common response for the other respondents. For the "educatn" variable all values equal to or greater than 98 are recoded as 12, which means that they completed 12th grade. Again, this was the most frequent response. Below is the code.

```
PSID$kids[PSID$kids >= 98] <- 3
PSID$educatn[PSID$educatn >= 98] <- 12
```

Another peek at the histograms for these two variables and things look much better.

hist(PSID$kids)

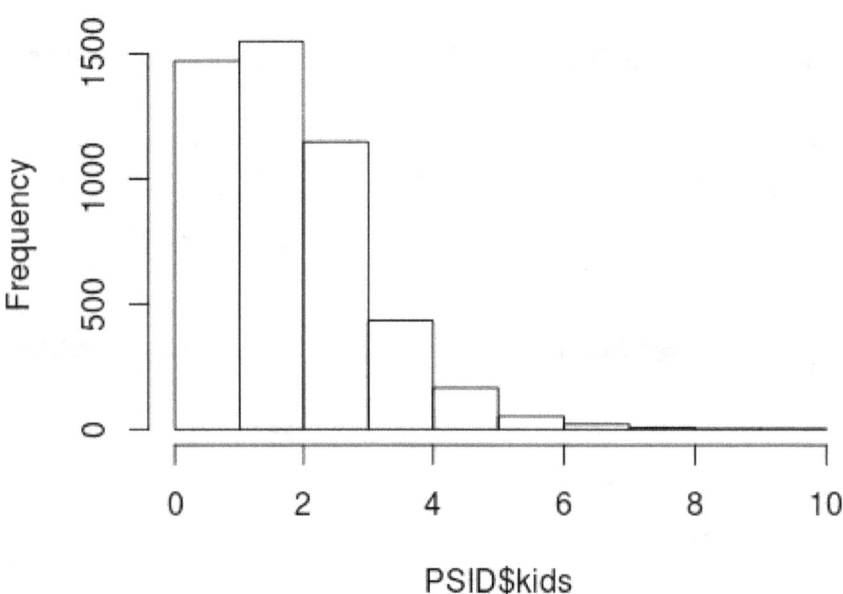

Figure 2.2: Histogram of Kids

hist(PSID$educatn)

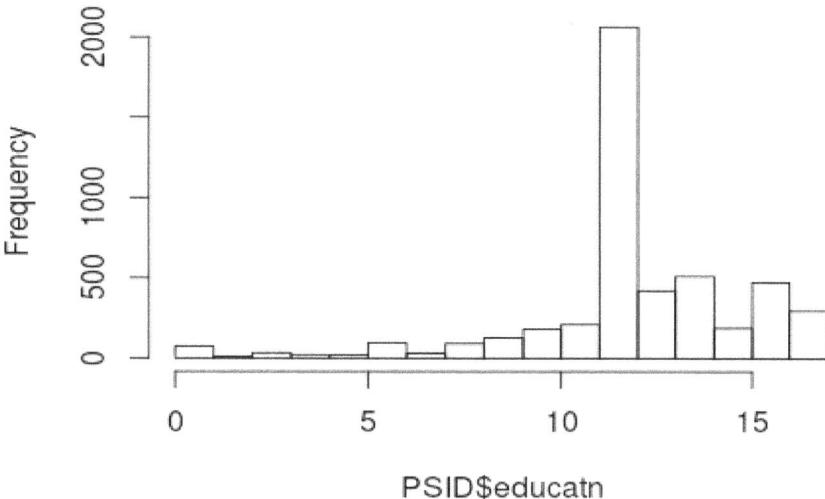

Figure 2.3: Histogram of Education

Now that everything is cleaned up we now need to make are training and testing data sets. This time we will do a manual split. This is because the data has no order to it. This also serves as another way to split your data. The code is below.

```
PSID_train<-PSID[1:3642,]
PSID_test<-PSID[3643:4856,]
```

Model Development

We will now make our model and create a visual of it. Our goal is to predict the number of children a person has based on their age, education, earnings, hours worked, and marital status. In the last chapter, we did our analysis using the "caret" package and create the visual with the "rattle" package. This time we will do our analysis by using the "rpart" package directly. We could do everything the same way every single time. That does make sense from a teaching perspective but the real world involves multiple ways to solve the same problem so this is why we are using different packages to complete the same task.

To perform the actual analysis we will use the "rpart" function to create the regression tree and the "rpart.plot" to create the visual. For the visual we indicate the number of digits at three, move all the leaves to the bottom by setting "fallen.leaves" argument to "TRUE", and use "extra" set to 101 to tell us how many examples in each leaf. Below is the code and figure 2.4 is the actual tree.

```
#make model
PSID_Model<-rpart(kids~age+educatn+earnings+hours+married, PSID_train)
#make visualization
rpart.plot(PSID_Model, digits=3, fallen.leaves = TRUE,type = 3, extra=101)
```

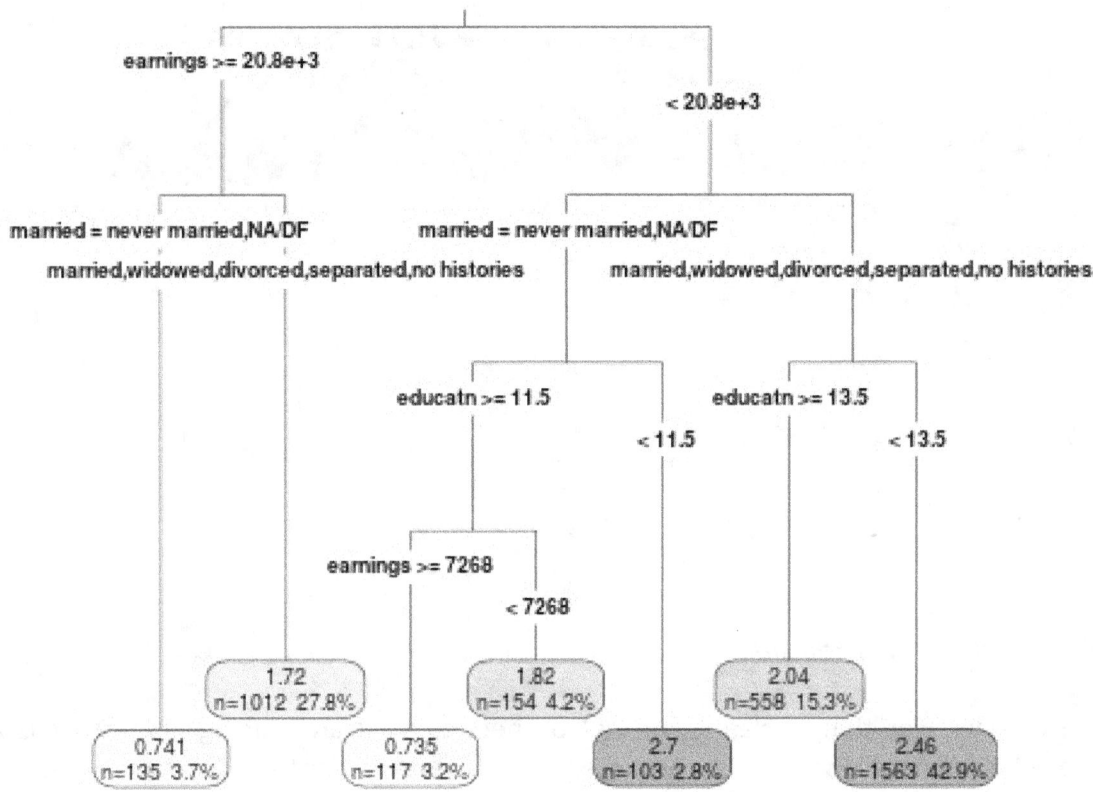

Figure 2.4 Regression tree

      The first split on the tree is by income. On the left, we have those who make more than 20,800 and on the right those who make less than 20,800. On the left the next split is by marriage, those who are never married or not applicable have on average 0.74 kids and represent 3.7% of the sample. Those who are married, widowed, divorced, separated, or have no history have on average 1.72 and make 28% of the sample.

      The right side of the tree is much more complicated and I will not explain all of it. After making less than 20,800 the next split is by marriage. Those who are married, widowed, divorced, separated, or no history with less than 13.5 years of education have 2.46 children on average and consist of 43% of the sample.

Model Testing

Our next task is to make the prediction model. We will do this with the following code

`PSID_pred<-`**`predict`**`(PSID_Model, PSID_test)`

      We will now evaluate the model. We will do this three different ways.
1. The first involves looking at the summary statistics of the prediction model and the testing data. The numbers should be about the same.
2. After that, we will calculate the correlation between the prediction model and the testing data.

3. Lastly, we will use a technique called the mean absolute error. Below is the code for the summary statistics and correlation.

```
summary(PSID_pred)
##    Min. 1st Qu.  Median    Mean 3rd Qu.    Max.
##   0.735   2.041   2.463   2.226   2.463   2.699
summary(PSID_test$kids)
##    Min. 1st Qu.  Median    Mean 3rd Qu.    Max.
##   0.000   2.000   2.000   2.494   3.000  10.000
cor(PSID_pred, PSID_test$kids)
## [1] 0.308116
```

Looking at the summary stats our model has a hard time predicting extreme values because the max value of the two models are far apart. The model predicts that the max number of kids a person will have is 2.69 but the test data has a max of 10. However, how often do people have ten kids? As such, this is not a major concern. It might be wise to remove the people with ten kids as an outlier. The other stats are reasonable especially the mean and the first quartile.

However, a look at the correlation finds that it is pretty low (0.30) this means that the two models have little in common. A high correlation would mean that both models have a similar pattern. However, we have to keep in mind that this data is count data and so the correlation has to view with skepticism. As such, we may need to make some changes in order to improve our model if this was for an actual application.

The mean absolute error is a measure of the difference between the predicted and actual values in a model. We need to make a function first before we analyze our model.

```
MAE<-function(actual, predicted){
    mean(abs(actual-predicted))
}
```

We now assess the model with the code below

```
MAE(PSID_pred, PSID_test$kids)
## [1] 1.134968
```

The results indicate that on average the difference between our model's prediction of the number of kids and the actual number of kids was 1.13 on a scale of 0 – 10 (the minimum and maximum number of kids possible). That is a lot of error in this particular context because we can mis-predict how many kids a person has by over one, which is a larger error when counting children. However, we need to compare this number to how well the mean of our model does to give us a benchmark. The code is below.

```
ave_kids<-mean(PSID_train$kids)
MAE(ave_kids, PSID_test$kids)
## [1] 1.178909
```

Therefore, it appears that the MAE is not much better than the mean for the model. This is highly concerning because we develop models that we hope are superior to the model that is generated using the mean.

Modal Tree Development and Testing

Our regression tree model with a MAE of 1.13 is slightly better than using the mean, which is 1.17. We will try to improve our model by switching from a regression tree to a model tree, which uses a slightly different approach for prediction. In a modal tree, each node in the tree ends in a linear regression model. Below is the code followed by the results.

```
PSIDM5<- M5P(kids~age+educatn+earnings+hours+married, PSID_train)
PSIDM5
## M5 pruned model tree:
## (using smoothed linear models)
##
## earnings <= 20754 :
## |   earnings <= 2272 :
## |   |   educatn <= 12.5 : LM1 (702/111.555%)
## |   |   educatn >  12.5 : LM2 (283/92%)
## |   earnings >  2272 : LM3 (1509/88.566%)
## earnings >  20754 : LM4 (1147/82.329%)
##
## LM num: 1
## kids =
##   0.0385 * age
## + 0.0308 * educatn
## - 0 * earnings
## - 0 * hours
## + 0.0187 * married=married,divorced,widowed,separated,no histories
## + 0.2986 * married=divorced,widowed,separated,no histories
## + 0.0082 * married=widowed,separated,no histories
## + 0.0017 * married=separated,no histories
## + 0.7181
##
## LM num: 2
## kids =
##   0.002 * age
## - 0.0028 * educatn
## + 0.0002 * earnings
## - 0 * hours
## + 0.7854 * married=married,divorced,widowed,separated,no histories
## - 0.3437 * married=divorced,widowed,separated,no histories
## + 0.0154 * married=widowed,separated,no histories
## + 0.0017 * married=separated,no histories
## + 1.4075
##
## LM num: 3
## kids =
##   0.0305 * age
## - 0.1362 * educatn
```

```
##   - 0 * earnings
##   - 0 * hours
##   + 0.9028 * married=married,divorced,widowed,separated,no histories
##   + 0.2151 * married=widowed,separated,no histories
##   + 0.0017 * married=separated,no histories
##   + 2.0218
##
## LM num: 4
## kids =
##   0.0393 * age
##   - 0.0658 * educatn
##   - 0 * earnings
##   - 0 * hours
##   + 0.8845 * married=married,divorced,widowed,separated,no histories
##   + 0.3666 * married=widowed,separated,no histories
##   + 0.0037 * married=separated,no histories
##   + 0.4712
##
## Number of Rules : 4
```

It would take too much time to explain everything in this model. You can read part of this model as follows

- earnings greater than 20,754 use linear model 4. This is similar to the first split in our regression tree, which was at 20,800.
- earnings less than 20,754 and less than 2,272 and less than 12.5 years of education use linear model 1,
- earnings less than 20,754 and less than 2,272 and greater than 12.5 years of education use linear model 2
- earnings less than 20,754 and greater than 2,272 linear model 3

Next, the printout shows the equation for each of the linear models. What is happening is that an example is first classified. After being classified the model knows which regression equation to use to predict the number of children that example has.

Lastly, we will evaluate our modal tree using the same criteria as the regression tree model. Below is the code.

```
PSIDM5_Pred<-predict(PSIDM5, PSID_test)
summary(PSIDM5_Pred)
##    Min. 1st Qu.  Median    Mean 3rd Qu.    Max.
##  0.3654  2.0487  2.3405  2.3365  2.6862  4.4217
cor(PSIDM5_Pred, PSID_test$kids)
## [1] 0.3486492
MAE(PSID_test$kids, PSIDM5_Pred)
## [1] 1.088617
```

This model is slightly better. For example, the modal tree is better at predicting extreme values at 4.4 compare to 2.69 for the regression tree model. The correlation is 0.34 is better than 0.30 for the regression tree but not as strong as we would hope. Lastly, the mean absolute error shows a slight improve to 1.08 compared to 1.13 in the regression tree model. However, we still are falling short in terms of making a strong model.

The improvement in the modal tree is to be expected as they are usually more accurate. However, remember the problem with model trees is in the interpretation and explanation of them and not as much their performance.

Conclusion

This chapter explained the use and development of numeric prediction trees. Both regression and model trees are available for use for the data scientist. Regression trees focus on using the means at a give leaf while modal trees use regression at each leaf. Assessing these models involves comparisons of summary statistics, correlational analysis, as well as the calculation of error. In the next chapter, we will be exposed to a variation on decision tress called random forest.

Chapter Three: Random Forest

So far, we have been making one tree at a time. However, why make one tree if you can make several, perhaps hundreds of trees. The more the merrier as we like to say. In this chapter, we will look at random forest and its application in both classification and numeric prediction.

Chapter Objectives
The objectives of this chapter are as follows.
- Examine concerns with decision trees
- Define random forest
- Share the pros and cons of random forest
- Prepare data for random forest analysis
- Develop and test a random forest classification tree
- Develop and test a random forest regression tree

Concerns with Decision Trees
One problem with decision trees is that it is easy to overfit the model with only one tree. Overfitting is the situation in which the model performs so well on the training data that it cannot repeat this performance on other datasets. Models are only useful when they are consistent in what they predict. If prediction is unpredictable, (no pun intended) the model is useless.
　　One solution to overfitting is the development of an algorithm that develops multiple trees rather than a single tree. This reduces the chances for overfitting to occur by boosting the stability of the estimates.

Random Forest Explained
Random forest involves the process of creating multiple decision trees and then combining the results. How this is done is through r using 2/3 of the data set to develop a decision tree. This is done dozens, hundreds, or more times. Every tree made is created with a slightly different sample. This process of sampling is called bootstrap aggregation or bagging for short. The results of all these trees are then averaged together if it is numeric. If it is classification than the different trees all vote on the outcome with simply majority the default setting for who wins. For example, if I am trying to predict gender and I make 3 trees. If 2 trees predict male and one predicts female the score is 2-1 and the final prediction

would be male. This process of using trees to decide is the basis for ensemble learning in the realm of machine learning. Here multiple trees are learning together on slightly different datasets to produce much more accurate results.

While the random forest algorithm is developing different samples it also randomly selects which variables to use for each tree that is developed. By randomizing the sample and the features used in the tree, random forest is able to reduce both bias (underfitting) and variance (overfitting) in a model. In addition, random forest is robust against outliers and collinearity.

Pros and Cons of Random Forest
The pros of random forest have already been discussed. You are able to reduce overfitting and improve performance through the use of a multitude of trees rather than one tree.

One problem with random forest is visualizing the data. It would not be practical to plot hundreds of trees. As such, for visual learners this can be a problem. Lastly, it is generally difficult to interpret the output as well.

Data Preparation
In this chapter, we will look at an application of random forest in R using both classification and numeric prediction. For classification, we will use the 'College' data from the 'ISLR' package to predict whether a college is private or not.

After loading our packages and setting our data to "College" we need to split our data into a training and testing. We will use the "createDataPartition" function form the "caret" package to do this. Next, we will create training and testing sets through subsetting them from the "College" dataset. Lastly, we will print the available variables.

```
library(ggplot2);library(ISLR);library(caret);library(corrplot)
data("College")
forTrain<-createDataPartition(y=College$Private, p=0.7, list=FALSE)
trainingset<-College[forTrain, ]
testingset<-College[-forTrain, ]
str(College)
## 'data.frame':    777 obs. of  18 variables:
##  $ Private    : Factor w/ 2 levels "No","Yes": 2 2 2 2 2 2 2 2 2 2 ...
##  $ Apps       : num  1660 2186 1428 417 193 ...
##  $ Accept     : num  1232 1924 1097 349 146 ...
##  $ Enroll     : num  721 512 336 137 55 158 103 489 227 172 ...
##  $ Top10perc  : num  23 16 22 60 16 38 17 37 30 21 ...
##  $ Top25perc  : num  52 29 50 89 44 62 45 68 63 44 ...
##  $ F.Undergrad: num  2885 2683 1036 510 249 ...
##  $ P.Undergrad: num  537 1227 99 63 869 ...
##  $ Outstate   : num  7440 12280 11250 12960 7560 ...
##  $ Room.Board : num  3300 6450 3750 5450 4120 ...
##  $ Books      : num  450 750 400 450 800 500 500 450 300 660 ..
##  $ Personal   : num  2200 1500 1165 875 1500 ...
##  $ PhD        : num  70 29 53 92 76 67 90 89 79 40 ...
##  $ Terminal   : num  78 30 66 97 72 73 93 100 84 41 ...
##  $ S.F.Ratio  : num  18.1 12.2 12.9 7.7 11.9 9.4 11.5 13.7 11.3 11.5 ...
##  $ perc.alumni: num  12 16 30 37 2 11 26 37 23 15 ...
```

```
##  $ Expend     : num  7041 10527 8735 19016 10922 ...
##  $ Grad.Rate  : num  60 56 54 59 15 55 63 73 80 52 ...
```

We have 18 variables to choose from. We will not print the histograms here in order to save space. Below, Figure 3.1 is the correlation plot

```
corrplot(cor(College[,-1]),method = 'number',col='black')
```

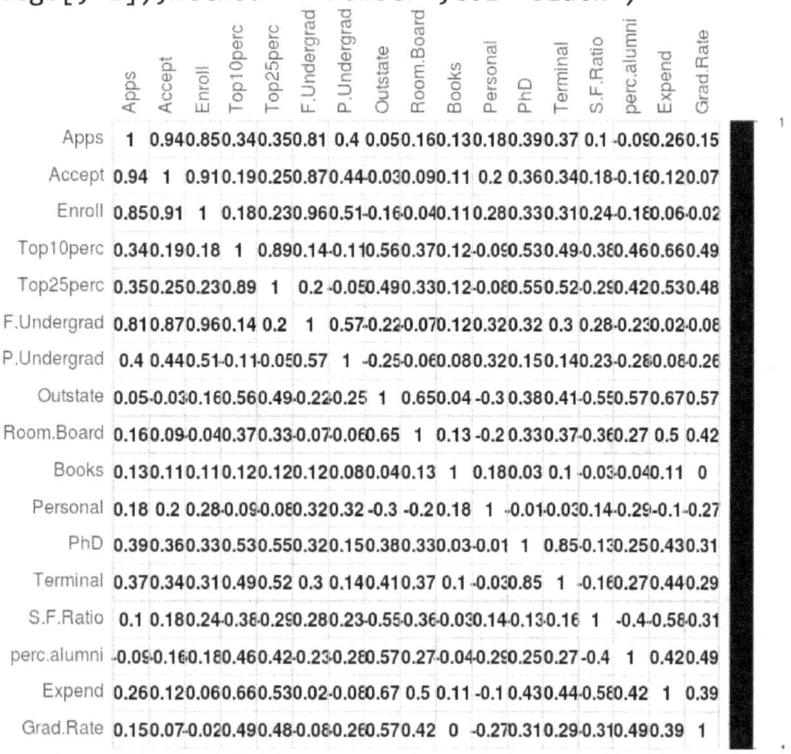

Figure 3.1: Correlation plot of "College"

     There are several high correlation in the plot. For simplicity, we will only use 7 variables to predict a colleges status as public or private.

Random Forest Classification
Model Development
Next, we need to setup the model we want to run using Random Forest. For this we will return to using the "caret" package and use the "train" function. Inside the function, we need to set the method to "rf" (random forest) and "prox" argument to "TRUE." Below is the code

```
Model1<-train(Private~Grad.Rate+Outstate+Room.Board+Books+PhD+S.F.Ratio+Expend,
data=trainingset, method='rf',prox=TRUE)
```

The initial output is indicated below

```
Model1
## Random Forest
##
## 545 samples
##   7 predictors
```

```
##    2 classes: 'No', 'Yes'
##
## No pre-processing
## Resampling: Bootstrapped (25 reps)
## Summary of sample sizes: 545, 545, 545, 545, 545, 545, ...
## Resampling results across tuning parameters:
##
##   mtry  Accuracy   Kappa
##   2     0.8942508  0.7280215
##   4     0.8944565  0.7306476
##   7     0.8944838  0.7310420
##
## Accuracy was used to select the optimal model using  the largest value.
## The final value used for the model was mtry = 7.
```

Most of this is self-explanatory and is a repeat of a chapter 1 except with different data. The tuning parameter for random forest is "mtry." Mtry is the number of variables randomly sampled as candidates at each split in the tree.

At the bottom of the output, the computer tells which mtry was the best. For our example, the best mtry was number 7. If you look closely, you will see that mtry 7 had the highest accuracy and Kappa as well.

Model Testing

We will now use the testing data to check the accuracy of the model we developed on the training data. Below is the code followed by the output.

```
pred <- predict(Model1, testingset)
testingset$predRight<-pred==testingset$Private
table(pred, testingset$Private)
##
## pred   No  Yes
##   No   52   8
##   Yes  11  161
mean(predNew==testingset$Private)
## [1] 0.9181034
```

For the most part, the model we developed to predict if a university is private or not is highly accurate at 91%.

Another form of useful analysis with random forest is to determine which variables were most useful in the predictions. You can calculate how important an individual variable is in the model. To find out the importance of each variable we will use the "varImp" function from the "caret" package and we will than plot this as show in figure 3.2

```
varImp(Model1)
## rf variable importance
##
##              Overall
## Outstate     100.00
## S.F.Ratio     57.65
```

```
## PhD            27.24
## Expend         22.33
## Room.Board     13.95
## Grad.Rate      11.85
## Books           0.00
plot(varImp(Model1))
```

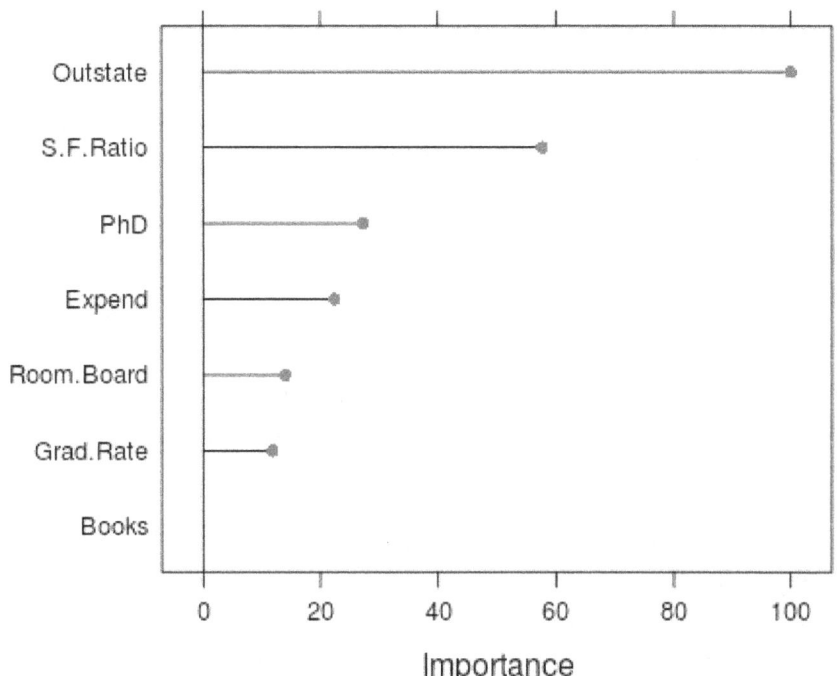

Figure 3.2: Variable importance plot

From the plot, it is clear that the variable "Outstate" was the most important in determining if a college was private or public. This was followed by "S.F.Ratio" or student to faculty ratio.

Random Forest Regression

Below is an example of random forest using regression trees. We are using the same data. The only difference is that we want to predict enrollment. Since enrollment is numeric, we are now conducting a random forest regression analysis

```
library(ggplot2);library(ISLR);library(caret)
data("College")
forTrain<-createDataPartition(y=College$Private, p=0.7, list=FALSE)
trainingset<-College[forTrain, ]
testingset<-College[-forTrain, ]
```

Model Development

The model is that same except for the prediction of enrollment. We also added the variable "Private" as a predictor. Lastly, notice how the "importance" argument is set to TRUE. This will be important when we look at the variable importance in the model.

```
model_enroll<-
train(Enroll~Grad.Rate+Outstate+Room.Board+Books+PhD+S.F.Ratio+Expend+Private,
data=trainingset, method='rf',prox=TRUE,importance=TRUE)
model_enroll
## Random Forest
##
## 545 samples
##   8 predictors
##
## No pre-processing
## Resampling: Bootstrapped (25 reps)
## Summary of sample sizes: 545, 545, 545, 545, 545, 545, ...
## Resampling results across tuning parameters:
##
##   mtry  RMSE       Rsquared
##   2     678.6387   0.5412901
##   5     684.4705   0.5282241
##   8     700.3526   0.5129319
##
## RMSE was used to select the optimal model using  the smallest value.
## The final value used for the model was mtry = 2.
```

This output is different from classification. There is no mention of accuracy or kappa because that only applies to classification. Instead, you have root mean squared error (RMSE) and the r squared along with a repeat of mtry.

RMSE and r-square should be familiar if you are acquainted with regression. At the bottom of the output, the computer tells which mtry was the best. For our example, the best mtry was number 2. Naturally, this model has the lowest error and highest r-square.

Model Testing

We will now use the testing data to check the accuracy of the model we developed on the training data. First we will create the prediction model. Then, we will calculate the correlation between the predicted values and the actual values in the testing set. Next, we will look at the summary statistics. Finally, we will calculate the error. Below is the code followed by the output

```
pred_enroll <- predict(model_enroll, testingset)
cor(pred_enroll, testingset$Enroll)
## [1] 0.7259437

summary(testingset$Enroll)
   Min. 1st Qu.  Median   Mean 3rd Qu.    Max.
   35.0   243.5   441.0  779.9   919.2  4615.0
summary(pred_enroll)
   Min. 1st Qu.  Median   Mean 3rd Qu.    Max.
  156.4   371.6   520.1  733.6   961.9  2929.5

MAE<-function(actual, predicted){
        mean(abs(actual-predicted))
}
MAE(pred_enroll, testingset$Enroll)
```

[1] 642.7268

The correlation is strong so things look great. However, there are some warning signs. For example, in the summary statistics the model has a problem predicting extreme values. In addition, a MAE of 642 is a lot of error when you consider the minimum and maximum enrollment in the testing set (35 -- 4,615). In this situation, since the correlation is high but so is the error, it is useful to make a plot of the prediction model and actual enrollment figures. Figure 3.3 is a plot of the model and actual results.

plot(pred_enroll, testingset$Enroll)

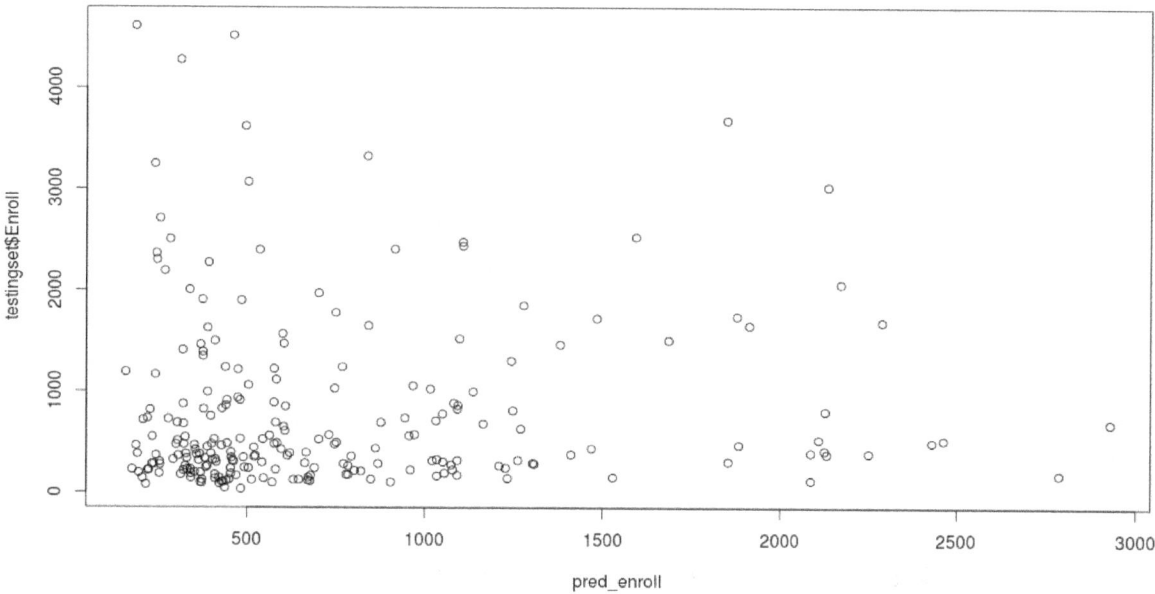

Figure 3.3: Model vs Actual Results

With the visual, you can clearly see the model is not doing a great job at predicting. We will now calculate how important an individual variable is in the model. As a note, you can only calculate the variable importance with the model you created from the training data and not the prediction model. After determining variable importance there is a plot of it in figure 3.4.

varImp(model_enroll)
```
## rf variable importance
##
##              Overall
## PrivateYes   100.00
## PhD           60.56
## S.F.Ratio     45.49
## Outstate      45.16
## Expend        40.89
## Room.Board    22.49
## Grad.Rate     22.08
## Books          0.00
```

```
plot(varImp(model_enroll))
```

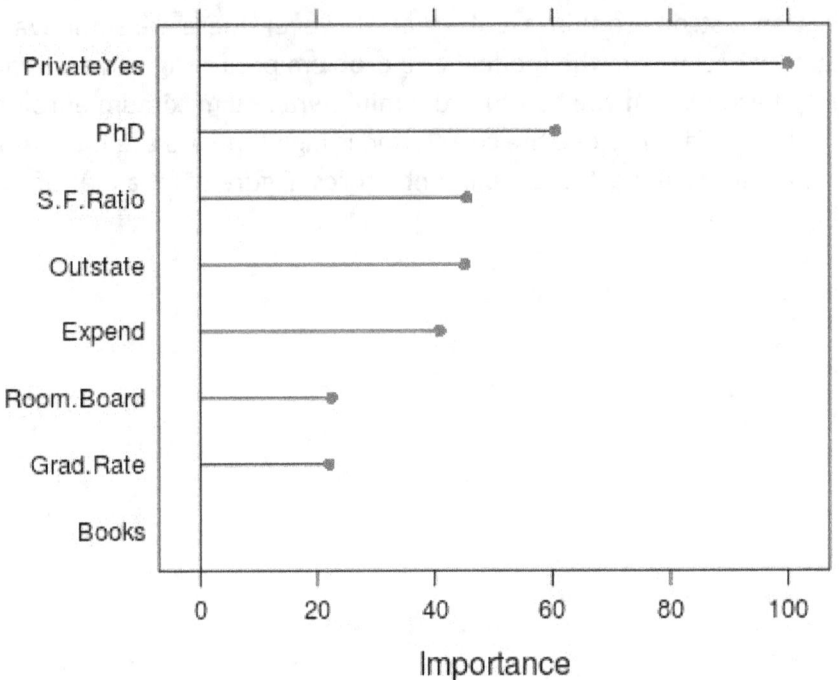

Figure 3.4: Variable importance for enrollment

The strongest factor in enroll appears to be whether the college is private or not followed by the proportion of PhDs.

Conclusion
In this chapter, we looked at random forest and its use. Random forest allows you to make a countless number of decision trees whether classification or numeric. This reduces instability and overfitting through the complex sampling that it does. We will now turn our attention to classification rules.

Chapter Four: Classification Rules

Classification rules are an alternative to decision trees in many contexts. They are somewhat like decision trees but have several distinct characteristics that we will explore.

Chapter Objectives
The objectives of this chapter are as follows.
- Define Classification Rules
- Explain the algorithms used to make classification rules
- Share the pros and cons of Classification Rules
- Prepare data for analysis
- Develop and test a random forest regression tree

Classification Rules
Classification rules represent knowledge in an if-else format. If you are familiar with computer programming than you are familiar with if-else statements. Basically, an if-else statements means "if this happens do "A" or else if this does not happen do "B." There are two parts to a classification rule. These two parts are the antecedent and consequent. Antecedent is the before aspect of the rule and consequent is after a given event takes place. An example of a classification rule is as follows...
- **If** the students studies at least 5 hours a week **then** they will pass the class with an A or **else** they will not pass with an A

This simple rule can be broken down into the following antecedent and consequent.
- Antecedent–If the student studies at least 5 hours a week
- Consequent-then they will pass the class with an A

The antecedent determines if the consequent takes place. For example, the student must study at least 5 hours a week to get an A. This is the rule in this particular context. Anything less than 5 hours and the students will not get an A.

Classification rules are developed on current data to make decisions about future actions. They are highly similar to the more common decision trees. The primary difference is that decision trees involve a complex systematic process to make a decision.

The Algorithms

Classification rules use algorithms that employ a separate and conquer heuristic like decision trees. What this means is that the algorithm will try to separate the data into smaller and smaller subset by generating enough rules to make homogeneous subsets. The goal is always to separate the examples in the data set into subgroups that have similar characteristics.

Common algorithms used in classification rules include the One Rule Algorithm and the RIPPER Algorithm. The One Rule Algorithm analyzes data and generates one all-encompassing rule. This algorithm works by finding the single rule that contains the least amount of error. Despite its simplicity, it is surprisingly accurate and it may be the easiest model to explain in all of machine learning.

The RIPPER algorithm grows as many rules as possible. When a rule begins to become so complex that it no longer helps to purify the various groups the rule is pruned or the part of the rule that is not beneficial is removed. This process of growing and pruning rules is continued until there is no further benefit.

RIPPER algorithm rules are more complex than One Rule Algorithm. This allows for the development of complex models. The drawback is that the rules can become too complex to make practical sense.

Pros and Cons of Classification Rules

Classification rules are stand-alone rules that are abstracted from a process. This means that if the results make several rules you can only use one of them in future application with some success. This is not possible with a decision tree. A decision tree is integrated and you must take the entire model with you for any future applications of it.

Another strength of classification rules is that you do not need to be familiar with the process that created it. While with decision trees, you do need to be familiar with the process that generated the decision.

One catch with classification rules in machine learning is that the majority of the variables need to be nominal in nature. As such, classification rules are not as useful for a model that may need to include a large number of continuous variables. This is not a problem with decision trees. The reason being is that if you have many continuous variables the classification algorithm will divide the continuous variables into dozens of sub-categories that lack any meaning. As you know, continuous variables can be divided an infinite number of ways and this can cause problems in interpreting the results. Also classification rules can only classify so numeric prediction is not possible.

Just as a note, decision trees can make rules but they are complex at times. However, for computational reasons, there are times when using decision trees to generate rules is appropriate but this is beyond the scope of this text.

Data Preparation

We are now going to analyze some data in order to make some classification rules. Specifically we are going to look at the "Males" dataset from the "Ecdat" package. Our goal is to make some rules that classify a male as married or not. Below is the code to load the needed packages as well as the dataset "Males"

```
library(Ecdat);library(RWeka);library(caret)
data("Males")
str(Males)
```

```
## 'data.frame':    4360 obs. of  12 variables:
##  $ nr        : int  13 13 13 13 13 13 13 13 17 17 ...
##  $ year      : int  1980 1981 1982 1983 1984 1985 1986 1987 1980 1981 ...
##  $ school    : int  14 14 14 14 14 14 14 14 13 13 ...
##  $ exper     : int  1 2 3 4 5 6 7 8 4 5 ...
##  $ union     : Factor w/ 2 levels "no","yes": 1 2 1 1 1 1 1 1 1 1 ...
##  $ ethn      : Factor w/ 3 levels "other","black",..: 1 1 1 1 1 1 1 1 1 1 ...
##  $ maried    : Factor w/ 2 levels "no","yes": 1 1 1 1 1 1 1 1 1 1 ...
##  $ health    : Factor w/ 2 levels "no","yes": 1 1 1 1 1 1 1 1 1 1 ...
##  $ wage      : num  1.2 1.85 1.34 1.43 1.57 ...
##  $ industry  : Factor w/ 12 levels "Agricultural",..: 7 8 7 7 8 7 7 7 4 4 ...
##  $ occupation: Factor w/ 9 levels "Professional, Technical_and_kindred",..: 9 9 9 9 5 2 2 2 2 2 ...
##  $ residence : Factor w/ 4 levels "rural_area","north_east",..: 2 2 2 2 2 2 2 2 2 2 ...
```

The first two variables "nr" and "year" are not to useful for us. "nr" is an identification number and "year" is the year the data was collected. We should not expect much change in marriage rates over a few years and the identification number has no meaning in our analysis. Therefore, we will ignore these two variables.

We will use the "createDataPartition" function to create the training and testing sets.

```
inTrain<-createDataPartition(y=Males$maried,p=0.7, list=FALSE)
trainingset <- Males[inTrain, ]
testingset <- Males[-inTrain, ]
```

Next, we visualize the data with some tables and histograms. Integer variables will be visualized with histograms and factor variables with tables. The code and results are below.

```
prop.table(table(trainingset$maried))
## 
##        no       yes
## 0.5610875 0.4389125
prop.table(table(trainingset$union))
## 
##        no       yes
## 0.7513921 0.2486079
prop.table(table(trainingset$ethn))
## 
##     other     black      hisp
## 0.7232231 0.1228300 0.1539469
prop.table(table(trainingset$industry))
## 
##                   Agricultural                        Mining
##                     0.03144448                    0.01506715
##                   Construction                         Trade
##                     0.07402555                    0.27251883
##                 Transportation                       Finance
##                     0.06714707                    0.04028824
##      Business_and_Repair_Service              Personal_Service
```

```
##                        0.07107763                          0.01768752
##                       Entertainment                       Manufacturing
##                        0.01604979                          0.27481166
## Professional_and_Related Service              Public_Administration
##                        0.07730102                          0.04258107
```
```r
prop.table(table(trainingset$health))
```
```
##
##         no        yes
## 0.98395021 0.01604979
```
```r
prop.table(table(trainingset$residence))
```
```
##
##     rural_area     north_east  nothern_central         south
##      0.0274223      0.2362888        0.3244973     0.4117916
```
```r
prop.table(table(trainingset$occupation))
```
```
##
## Professional, Technical_and_kindred  Managers, Officials_and_Proprietors
##                          0.10383230                           0.09269571
##                        Sales_Workers                  Clerical_and_kindred
##                          0.05142483                           0.11136587
##        Craftsmen, Foremen_and_kindred               Operatives_and_kindred
##                          0.21028497                           0.19620046
##                 Laborers_and_farmers             Farm_Laborers_and_Foreman
##                          0.09891910                           0.01277432
##                      Service_Workers
##                          0.12250246
```
```r
hist(trainingset$school)
```

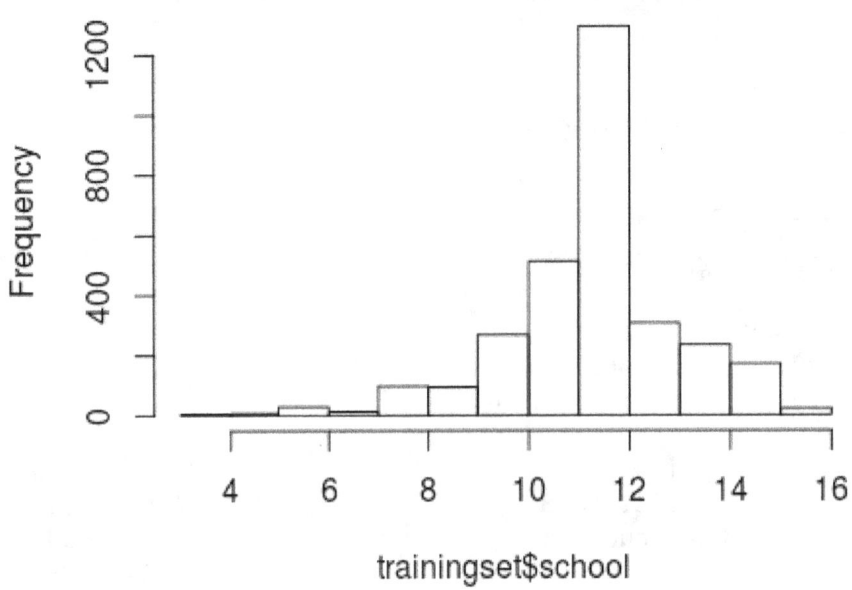

Figure 4.1: School histogram

`hist(trainingset$exper)`

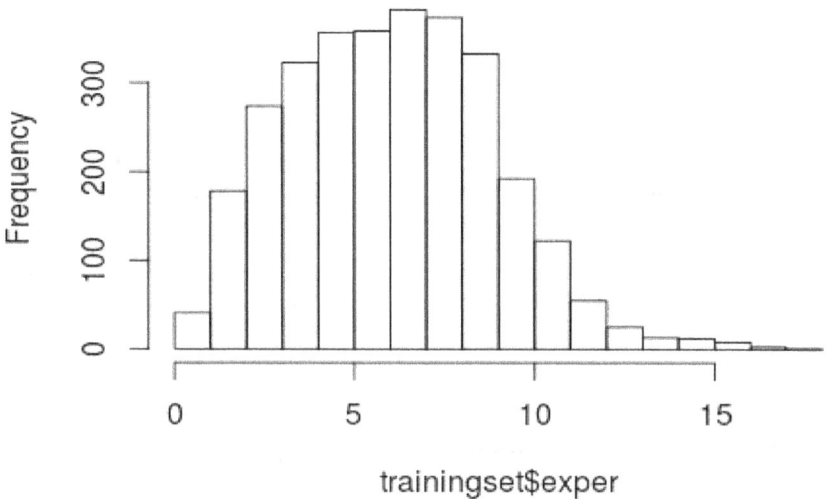

Figure 4.2: Exper histogram

`hist(trainingset$wage)`

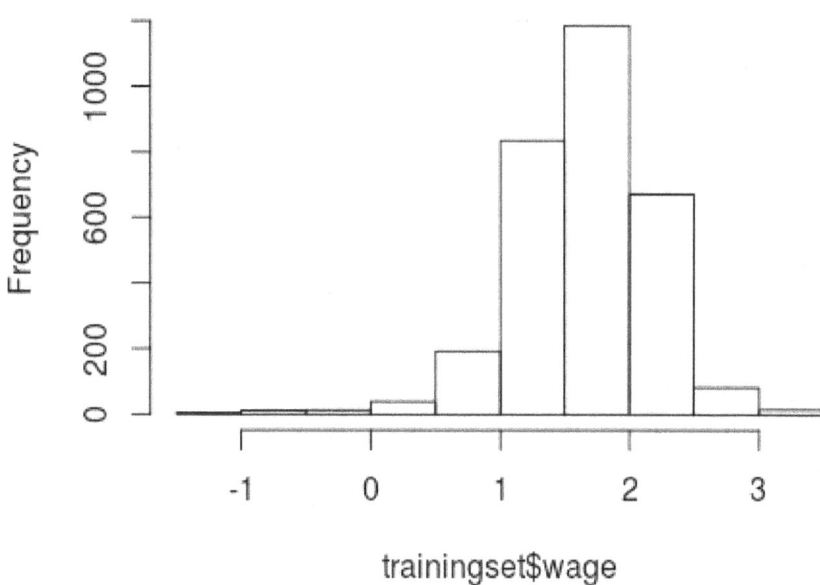

Figure 4.3: School histogram

There is no time or space to explain the tables and histograms in detail. Only two things are worth mentioning.
1. The outcome variable "maried" is mostly balance between those who are married and those who are not (56% to 44%).
2. The "health" variable is horribly unbalanced and needs to be removed (98% no vs 2% yes). As such, it will be remove from the model.

One Rule Algorithm
Model Development
We can now train our first model. The first model will be a one-rule model, which means that R will develop one all-encompassing rule for classification purposes. Below is the code.

```
set.seed(1)
Males_1R<-OneR(maried~ethn+union+industry+school+exper+occupation+residence,
data=trainingset)
```

We used the "OneR" function to create the model. This function analyzes the data and makes a single rule for it. We will now evaluate the model be first looking at the rule that was generated.

```
Males_1R
## exper:
##  < 7.5     -> no
##  >= 7.5    -> yes
## (1386/2188 instances correct)
```

The "exper" variable was selected for generating the rule. To state the rule clearly it literally it means
- IF a man has less than 7.5 years of experience he is not married ELSE
  If he has more than 7.5 years of experience THEN he is married.

Perhaps you can see the if-else statement. However, as we look at the accuracy of the model we will see some problems as you will notice below after typing in the following code

```
summary(Males_1R)
##
## === Summary ===
##
## Correctly Classified Instances        1386              63.3455 %
## Incorrectly Classified Instances       802              36.6545 %
## Kappa statistic                        0.229
## Mean absolute error                    0.3665
## Root mean squared error                0.6054
## Relative absolute error               75.3236 %
## Root relative squared error          122.74   %
## Total Number of Instances             2188
##
## === Confusion Matrix ===
##
```

```
##     a   b   <-- classified as
##   947 326 |   a = no
##   476 439 |   b = yes
```

We only correctly classified 63% of the data. This is pretty bad but for only one rule that is pretty impressive.

Model Testing
Below are the results with the test data

```
set.seed(1)
testmodel<-predict(Males_1R, newdata = testingset)
table(testmodel,testingset$maried)
##
## testmodel  no  yes
##       no  519  298
##       yes 214  276
(519+276)/(519+276+298+214)
## [1] 0.6082632
```

The one strength of our model is that it is consistently poor. The stability between training and testing is good but the overall results are bad. Perhaps if we change our approach and develop more than one rule we will have more success.

JRip Algorithm
Model Development
We will now use the "JRip" function to develop multiple classification rules. Below is the code.

```
set.seed(1)
Males_JRip<-JRip(maried~ethn+union+school+exper+industry+occupation+residence,
data=trainingset)
Males_JRip
## JRIP rules:
## ===========
##
## (exper >= 9) => maried=yes (491.0/198.0)
## (exper >= 5) and (occupation = Craftsmen, Foremen_and_kindred) and (ethn = other) => maried=yes (167.0/63.0)
## (exper >= 5) and (school >= 14) => maried=yes (152.0/62.0)
##  => maried=no (1378.0/428.0)
##
## Number of Rules : 4
```

There are four rules all together below is there meaning
- IF a male has at least 9 years or more of experience, THEN he is married. ELSE
- IF a male has at least 5 years or more of experience, his occupation is craftsmen or Foreman, his ethnicity is other THEN he is married. ELSE

- IF a male has at least 5 years or more of experience AND has at least 14 years of school or more THEN he is married. ELSE
- IF none of these rules apply THEN he is not married

The numbers in parentheses represent how many examples are covered by this rule as well as the number of misclassifications. For example, the top rule has 491 examples classified correctly and 198 examples classified incorrectly.

Notice how all rules begin with "exper" this is one reason why the "OneR" function made its rule on experience. Experience is the best predictor of marriage in this dataset. However, our accuracy has not improved much as you will see in the following code.

```
summary(Males_JRip)
##
## === Summary ===
##
## Correctly Classified Instances         1437               65.6764 %
## Incorrectly Classified Instances        751               34.3236 %
## Kappa statistic                           0.2831
## Mean absolute error                       0.4471
## Root mean squared error                   0.4728
## Relative absolute error                  91.8843 %
## Root relative squared error              95.8575 %
## Total Number of Instances              2188
##
## === Confusion Matrix ===
##
##     a    b   <-- classified as
##   950  323 |   a = no
##   428  487 |   b = yes
```

We are only at 65% which is not much better. Since this is a demonstration the actually numbers do not matter as much.

Modeling Testing
Below are the results on the test set

```
set.seed(1)
testmodel<-predict(Males_JRip, newdata = testingset)
table(testmodel,testingset$maried)
##
## testmodel  no yes
##       no  508 275
##       yes 225 299
(508+299)/(508+299+275+225)
## [1] 0.6174445
```

There is almost no improvement. This means that we would need to do something else besides switching algorithms in order to improve the model. If we had to choose one of these two models the one-rule model is probably better because the accuracy is about the same with only one rule. The simpler is almost always the preferred model in statistics.

Conclusion

Classification rules provide easy to understand rules for organizing data homogeneously. There are several algorithms available such as the One Rule and J ripper. This yet another way to analyze data with machine learning approaches. In the next chapter, we will begin our exploration of "black box methods" starting with Support Vector Machines.

Chapter Five: Support Vector Machines

Support vector machines (SVM) is one of those mysterious black box methods in machine learning. What this means is that it is hard to explain how the algorithm makes predictions even though we all know that they work. In this chapter, we will try to explain in simple terms what SVM are and their strengths and weaknesses.

Chapter Objectives
The objectives of this chapter are as follows.
- Define Support Vector Machines
- Share the pros and cons of Support Vector Machines
- Prepare data for analysis
- Develop and test a Support Vector Machines
- Compare models using different kernels

Defining Support Vector Machines
SVM is a combination of the algorithms nearest neighbor and linear regression. For the nearest neighbor characteristic of SVM, SVM uses the traits of an identified example to classify an unidentified one. For the regression aspect of SVM, SVM draws a line that divides the various groups in the data. It is preferred that the line is straight (as in regression) but this is not always the case.

This combination of using the nearest neighbor along with the development of a regression line leads to the development of what is called a hyperplane. The hyperplane is drawn in a place that creates the greatest amount of distance among the various groups identified in the data set. The examples in each group that are closest to the hyperplane are called "support vectors." They support the vectors by providing the boundaries between the various groups. This is why this algorithm is called support vector machines.

If for whatever reason a line cannot be straight because the boundaries are not nice and tight. R will still draw a straight line but make accommodations using a slack variable, which allows for error and or for examples to be in the wrong group.

Another tool used in SVM analysis is the kernel trick. A kernel will add a new dimension or feature to the analysis by combining features that were measured in the data. For example, latitude and longitude might be combine mathematically to make altitude. This new feature is now used to develop hyperplane(s) for the data that are more accurate in classifying the examples.

There are several different types of kernel tricks that achieve their goal using various mathematics. There is no rule for which one to use and employing different choices is the only strategy currently.

Pros and Cons of Support Vector Machines

The pros of SVM is their flexibility of use as they can be used to predict numbers or classify. SVM are also able to deal with noisy data and are easier to use than their sister black box method artificial neural networks. Lastly, SVM are often able to resist overfitting and are usually highly accurate.

Cons of SVM include they are still complex as they are a member of black box machine learning methods even if they are simpler than artificial neural networks. In addition, the lack of criteria over kernel selection makes it difficult to determine which model is the best.

Data Preparation

We are now going to predict whether or not someone has diabetes, by using SVM for classification purposes. To do this analysis you will need to use the 'kernlab' package and the "MASS" package. The "MASS" package contains two datasets for diabetes the "Pima.tr" and the "Pima.te". The people who made the dataset divided it for us into a training and test set. As such, we do not need to partition the data ourselves.

```
library(kernlab);library(MASS);library(corrplot)
data("Pima.tr")
data("Pima.te")
#split data
trainingset<-Pima.tr
testingset<-Pima.te
#explore data
str(Pima.tr)
## 'data.frame':    200 obs. of  8 variables:
##  $ npreg: int  5 7 5 0 0 5 3 1 3 2 ...
##  $ glu  : int  86 195 77 165 107 97 83 193 142 128 ...
##  $ bp   : int  68 70 82 76 60 76 58 50 80 78 ...
##  $ skin : int  28 33 41 43 25 27 31 16 15 37 ...
##  $ bmi  : num  30.2 25.1 35.8 47.9 26.4 35.6 34.3 25.9 32.4 43.3 ...
##  $ ped  : num  0.364 0.163 0.156 0.259 0.133 ...
##  $ age  : int  24 55 35 26 23 52 25 24 63 31 ...
##  $ type : Factor w/ 2 levels "No","Yes": 1 2 1 1 1 2 1 1 1 2 ...
```

It's always good to explore our data a little. Therefore, we will look at the histograms, tables, and the correlations of the variables. We will set the number of graphs per page to four in order to save space using the "par" function and the "mfrow" argument. Figure 5.1 has several histograms.

```
par(mfrow=c(2,2))
hist(Pima.tr$npreg)
hist(Pima.tr$glu)
hist(Pima.tr$bp)
hist(Pima.tr$skin)
```

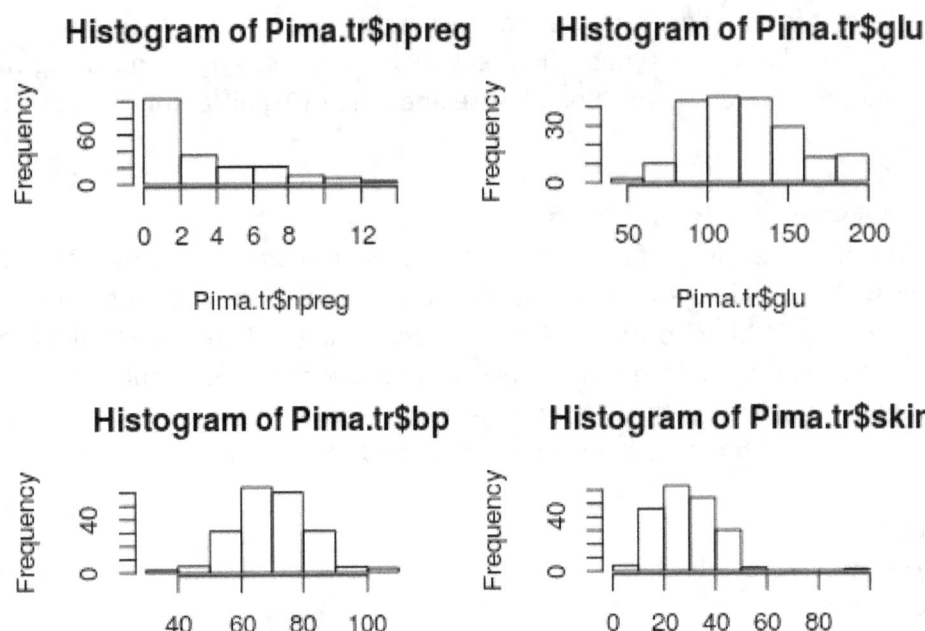

Figure 5.1: Histograms 1

```
hist(Pima.tr$bmi)
hist(Pima.tr$ped)
hist(Pima.tr$age)
table(Pima.tr$type)
## 
##  No Yes
## 132  68
```

"npreg" stands for the number of pregnancies a women has had. It would be expected for this to be non-normal as this is count data. "glu" is glucose level which is normal distributed. "bp" stands for blood pressure "skin" is a measurement of skin behind the triceps. Both of these are mostly normal in appearance. Lastly, the table shows us the proportion of the people with and without diabetes. Figure 5.2 has the remaining histograms

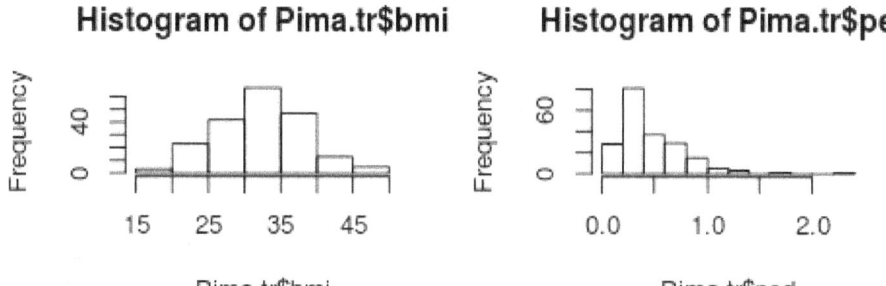
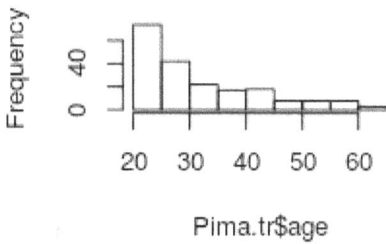

Figure 5.2 Histograms II

The remaining variables are somewhat non-normal. The "age" and the "ped" variables are non-normal. However, with the kernel tricks and other characteristics of SVM this should not be a serious problem.

Since regression is partially involved when using support vector machines we need to least look at the correlation among the variables. Before we run this code we will reset the number of graphs per page back to one and also remove the dependent variable because it is a factor and not numeric. Figure 5.3 is the correlation of the numeric variables.

```
par(mfrow=c(1,1))
corrplot(cor(Pima.tr[,-8]),method = 'number',col='black')
```

	npreg	glu	bp	skin	bmi	ped	age
npreg	1	0.17	0.25	0.11	0.06	-0.12	0.6
glu	0.17	1	0.27	0.22	0.22	0.06	0.34
bp	0.25	0.27	1	0.26	0.24	-0.05	0.39
skin	0.11	0.22	0.26	1	0.66	0.1	0.25
bmi	0.06	0.22	0.24	0.66	1	0.19	0.13
ped	-0.12	0.06	-0.05	0.1	0.19	1	-0.07
age	0.6	0.34	0.39	0.25	0.13	-0.07	1

Figure 5.3: Correlations

There appears that there are no major concerns with correlations either. Before we continue with the analysis, we need to scale our variables. This makes all variables to be within the same given range, which helps to equalize the influence of each of them on each other. This is necessary because SVM uses distances to make the groupings. Different scaling would mean that different variables would have different amounts of influence.

However, we do not want to change our "type" variable as this is the predictor variable and scaling it would make the results hard to understand and is technically impossible since we are classifying. Therefore, we are going to remove temporarily the "type" variable from both of our data sets and save them in a temporary data frame. The code is below.

```
#temporary dataframe for the label results
keep<-as.data.frame(trainingset$type)
keeptest<-as.data.frame(testingset$type)
#null label variable in both datasets
trainingset$type<-NA
testingset$type<-NA
```

Next, we scale the remaining variable and reinsert the label variables for each data set as show in our code below.

```
#scale datasets
trainingScaled<-as.data.frame(scale(trainingset))
testingScaled<-as.data.frame(scale(testingset))
#add back label
```

```
trainingScaled$type<-keep$`trainingset$type`
testingScaled$type<-keeptest$`testingset$type`
```

Linear Kernel
Model Development
Now we make our model using the "ksvm" function in the "kernlab" package. We set the kernel to "vanilladot" which is a linear kernel. We will also print the results. However, the results do not make any sense on their own and the model can only be assessed through other means. Below is the code.

```
#make the model
classify<-ksvm(type~., data=trainingScaled,
                kernel="vanilladot")
## Setting default kernel parameters
#look at the results
classify
## Support Vector Machine object of class "ksvm"
##
## SV type: C-svc  (classification)
##  parameter : cost C = 1
##
## Linear (vanilla) kernel function.
##
## Number of Support Vectors : 103
##
## Objective Function Value : -98.2129
## Training error : 0.23
```

The output tells you the type of SVM, which is classification. It also tells you the type of kernel used as well as the training error. The training error is .23 which means our accuracy is .77. There is also information on "cost C" we will discuss this idea in a later chapter.

Model Testing
We now need to use the "predict" function so that we can determine the accuracy of our model. Remember that for predicting, we use the answers in the test data and compare them to what our model would guess based on what it knows from the training set. In the code below, accuracy is calculated by determining how often the prediction of the model is the same as the answers in the testing set. This information is then displayed with the use of the "table" and "prop.table" functions.

```
predict_diabetes<-predict(classify, testingScaled)
table(predict_diabetes, testingScaled$type)
##
## predict_diabetes  No  Yes
##              No  198   42
##             Yes   25   67
accuracy<-predict_diabetes == testingScaled$type
prop.table(table(accuracy))
## accuracy
##     FALSE      TRUE
## 0.2018072 0.7981928
```

The table allows you to see how many examples were classified correctly and how they were misclassified. The "prop.table" function allows you to see this information as an overall percentage. This particular model was highly accurate at almost 80%. It would be difficult to improve further.

Radial Kernel
Model Development and Testing
Below is code for a model that is using a different kernel with results that are the same. The kernel is called "rbfdot" which means radial based function kernel. The details of the mathematics of this kernel is beyond the scope of this book.

```
classify_rbf<-ksvm(type~., data=trainingScaled,
                   kernel="rbfdot")
#evaluate improved model
classify_predict_rbf<-predict(classify_rbf, testingScaled)
table(classify_predict_rbf, testingScaled$type)
##
## classify_predict_rbf  No  Yes
##                 No   197   47
##                 Yes   26   62
accuracy_rbf<-classify_predict_rbf == testingScaled$type
prop.table(table(accuracy_rbf))
## accuracy_rbf
##     FALSE      TRUE
## 0.2198795 0.7801205
```

The results are essentially the same. Therefore, switching kernels did not provide any improvement in our predictions. If we want to improve the model, we could use other kernels are include other features in the model.

Conclusion
Support vector machines work even if it is not always clear why. Their application can be used in situations where the data is messy and accuracy is important. In the next chapter, we will see how support vector machines can be used with numeric prediction.

Chapter Six: Support Vector Machines Numeric Prediction

In this chapter, we will look at support vector machines for numeric prediction. As you know from the previous chapter, SVM is used for both classification and numeric prediction. One of the characteristics of SVM for numeric prediction or classification is that SVM will automatically create higher dimensions of the features and summarizes this in the output. In other words, unlike in regression or even in decision trees where you have to decide for yourself how to modify your features, SVM does this automatically using a kernel.

Chapter Objectives
The objectives of this chapter are as follows.
- Provide additional information about Support Vector Machines
- Prepare data for analysis
- Develop and test a Support Vector Machines
- Compare the performance of several Support Vector Machine kernels

More on SVM
Different kernels transform the features in different ways and we will look at several in this chapter. Something that we did not mention in the last chapter but will share here is that svm use what is called a cost function. The cost function determines the penalty for an example being on the wrong side of the margin developed by the kernel. The larger the cost function the fewer misclassifications are allowed.

Remember that SVM draws lines and separators to divide the examples in to homogeneous groups. Examples on the wrong side are penalized as determined by the researcher who sets the cost function. Just like with regression, generally the model with the least amount of error may be the best model when developing a numeric prediction svm.

Data Preparation

The purpose of this chapter is to use SVM to predict income in the "Mroz" dataset from the "Ecdat" package. We will use several different kernels that will transform the features different ways and calculate the mean-squared error and other metrics to determine the most appropriate model.

One important note, we will be using the "e1071" package for the support vector machine function instead of the "kernlab" package because the coding for "e1071" is slightly more intuitive and the printout easier to understand. In addition, exposure to various packages is helpful for any data scientist. Below is some initial code.

```
library(e1071);library(corrplot);library(Ecdat)
data(Mroz)
str(Mroz)
## 'data.frame':    753 obs. of  18 variables:
## $ work       : Factor w/ 2 levels "yes","no": 2 2 2 2 2 2 2 2 2 2 ...
## $ hoursw     : int  1610 1656 1980 456 1568 2032 1440 1020 1458 1600 ...
## $ child6     : int  1 0 1 0 1 0 0 0 0 0 ...
## $ child618   : int  0 2 3 3 2 0 2 0 2 2 ...
## $ agew       : int  32 30 35 34 31 54 37 54 48 39 ...
## $ educw      : int  12 12 12 12 14 12 16 12 12 12 ...
## $ hearnw     : num  3.35 1.39 4.55 1.1 4.59 ...
## $ wagew      : num  2.65 2.65 4.04 3.25 3.6 4.7 5.95 9.98 0 4.15 ...
## $ hoursh     : int  2708 2310 3072 1920 2000 1040 2670 4120 1995 2100 ...
## $ ageh       : int  34 30 40 53 32 57 37 53 52 43 ...
## $ educh      : int  12 9 12 10 12 11 12 8 4 12 ...
## $ wageh      : num  4.03 8.44 3.58 3.54 10 ...
## $ income     : int  16310 21800 21040 7300 27300 19495 21152 18900 20405 20425
## $ educwm     : int  12 7 12 7 12 14 14 3 7 7 ...
## $ educwf     : int  7 7 7 7 14 7 7 3 7 7 ...
## $ unemprate  : num  5 11 5 5 9.5 7.5 5 5 3 5 ...
## $ city       : Factor w/ 2 levels "no","yes": 1 2 1 1 2 2 1 1 1 1 ...
## $ experience : int  14 5 15 6 7 33 11 35 24 21 ...
```

We need to place the factor variables next to each other as it helps in having to remove them when we need to scale the data. We must scale the data because SVM is based on distance when making calculations as we discussed in the last chapter. If there are different scales, the larger scale will have more influence on the results. Below is the code followed by the correlational chart in figure 6.1.

```
mroz.scale<-Mroz[,c(17,1,2,3,4,5,6,7,8,9,10,11,12,13,14,15,16,18)] #move factor
variables next to each other
mroz.scale<-as.data.frame(scale(mroz.scale[,c(-1,-2)])) #remove factor variables
before scaling
mroz.scale$city<-Mroz$city # add factor variable back into the dataset
mroz.scale$work<-Mroz$work # add factor variable back into the dataset
corrplot(cor(mroz.scale[,1:16]),method = "number",col = "black")
```

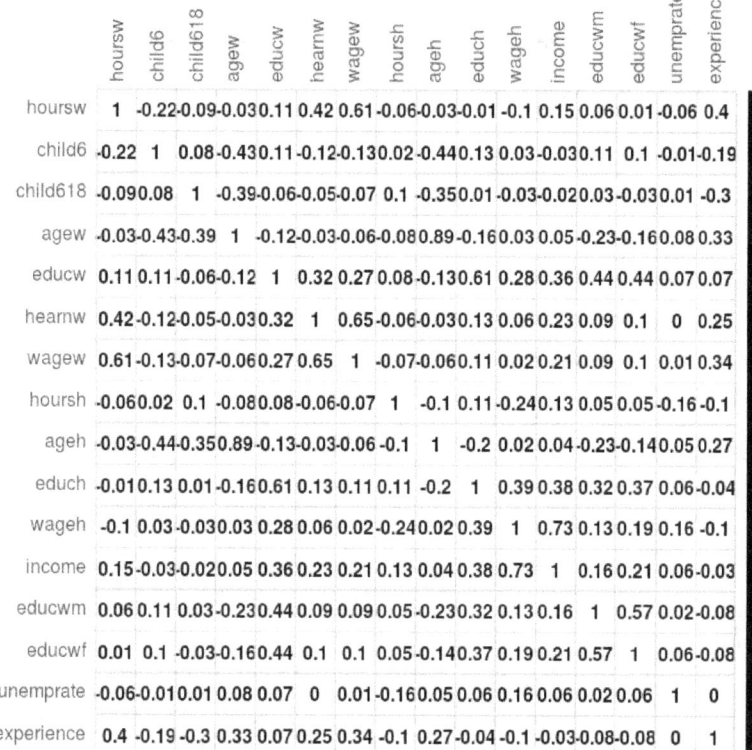

Figure 6.1: Correlational chart

There is a high correlation between "ageh" (age of husband) and "agew" (age of wife) but the rest is fine. For now, we will not make any changes.

Below is the code for creating the train and test datasets. This is a different approach from the "caret" package but it works the same as we are randomly assigning examples to each training set using the "sample" function. We are creating a variable called "ind" and assigning a 1 or 2 to every example. The size of the sample is the number of rows "nrow" in the dataset. We want to reuse our label of 1 or 2 until the dataset is completely labeled "replace = T". 70% of the dataset will be labeled a 1 and 30% labeled as a 2. After that, we subset our data based on the "ind" variable that we created. This is exactly what the "createDataPartition" function does in the "caret" package.

```
set.seed(502)
ind=sample(2,nrow(mroz.scale),replace=T,prob=c(.7,.3))
train<-mroz.scale[ind==1,]
test<-mroz.scale[ind==2,]
```

Model Development and Testing with Kernels
Our first kernel is the linear kernel. We are using the "tune.svm" function from the "e1071" package. We set the kernel to "linear" and we pick our own values for the cost function. Setting the values of the cost function is known as tuning the parameters. In a later chapter, we will explain this in greater detail.

In this chapter, we will tune the parameters using the "e1071" package with the "tune.svm" function. However, in a later chapter we will tune the parameters using the "caret" package. The process we are using to develop the models is as follows

1. Set the seed
2. Develop the initial model by setting the formula, dataset, kernel, and parameter(s)
3. Select the best model for the test set
4. Predict with the best model
5. Plot the predicted and actual results
6. Calculate the correlation and summary statistics
7. Calculate the mean squared error

The first time we will go through this process step-by-step. However, all future models will just have the code followed by an interpretation. Figure 6.2 provides a list of the kernels we will use as well as the parameters that need to be set for each kernel.

Kernel	Parameters
Linear	Cost function
Polynomial	Cost function, degree, coefficient theta
Radial	Cost function, gamma
Sigmoid	Cost function, gamma, Coefficient theta

Figure 6.2: Kernels and parameters

Linear Kernel
The first model will use a linear kernel. This kernel does not transform the data at all and is similar to regression. The numbers for the cost function can be whatever you want, however, it is common to set the cost function to cover a range of small and large increments as in the code below. Also keep in mind that r will produce six different models because we have six different values in the "cost" argument. Below is the code

```
linear.tune<-tune.svm(income~.,data=train,kernel="linear",cost =
c(.001,.01,.1,1,5,10))
summary(linear.tune)
##
## Parameter tuning of 'svm':
##
## - sampling method: 10-fold cross validation
##
## - best parameters:
##   cost
##    10
##
## - best performance: 0.3492453
##
## - Detailed performance results:
##    cost      error dispersion
## 1 1e-03 0.6793025  0.2285748
## 2 1e-02 0.3769298  0.1800839
## 3 1e-01 0.3500734  0.1626964
## 4 1e+00 0.3494828  0.1618478
## 5 5e+00 0.3493379  0.1611353
## 6 1e+01 0.3492453  0.1609774
```

The best model had a cost = 10 with a performance of .35. The lower the error the better is the numeric prediction just as in regression. We will now select the best model (cost =10) and use this on our test data. Below is the code.

```
best.linear<-linear.tune$best.model
tune.test<-predict(best.linear,newdata=test)
```

Now we will create a plot so we can see how well our model predicts as shown in figure 6.3. In addition, we will calculate the mean squared error to have an actually number of our models performance

```
plot(tune.test,test$income)
```

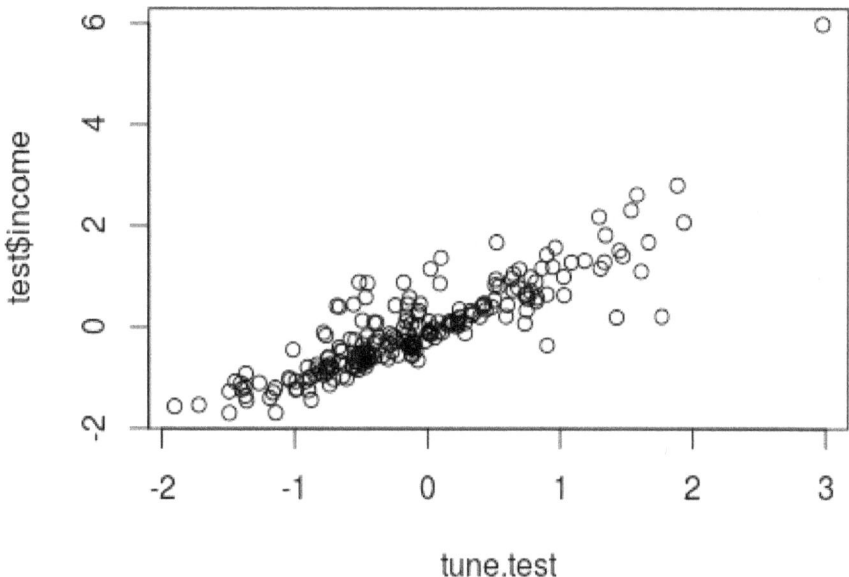

Figure6.3: Linear Kernel

This looks promising, below is the code to calculate the mean squared error, the correlation, and the summary statistics.

```
tune.test.resid<-tune.test-test$income
mean(tune.test.resid^2)
## [1] 0.215056
cor(tune.test,test$income)
##[1] 0.8721713
summary(tune.test)

    Min.   1st Qu.   Median      Mean   3rd Qu.     Max.
-1.90489  -0.60876  -0.15772  -0.09338   0.39465  2.97373
summary(test$income)
```

```
    Min.  1st Qu.   Median      Mean  3rd Qu.      Max.
-1.69649 -0.65500 -0.16247 -0.03755  0.44047   5.98180
```

The model looks good. However, we cannot tell if the error number is decent until it is compared to other models. The correlation is excellent and indicates that our current kernel models the data well. The summary statistics indicate that the model is mostly accurate but struggles with max values, however, this is not a major concern.

Before we move in on it is important to note that we will no longer adjust the cost function. The reason is that the other kernels already have several parameters to set. If we include the adjustment of the cost function in combination with adjusting the other parameters it would increase the computationally time a great deal. As such, for all future models the cost function is left at the default of 1 while we adjust the other parameters that are unique to each kernel.

Polynomial Kernel

The next kernel we will use is the polynomial one. The polynomial kernel is used for non-linear transformations for as many degrees as you specify. This kernel requires two parameters the degree of the polynomial (3,4,5,etc) as well as the kernel coefficient theta. Below is the code. Figure 6.4 is the polynomial kernel.

```
set.seed(123)
poly.tune<-tune.svm(income~.,data = train,kernal="polynomial",degree =
c(3,4,5),coef0 = c(.1,.5,1,2,3,4))
best.poly<-poly.tune$best.model
poly.test<-predict(best.poly,newdata=test)
plot(poly.test,test$income)
```

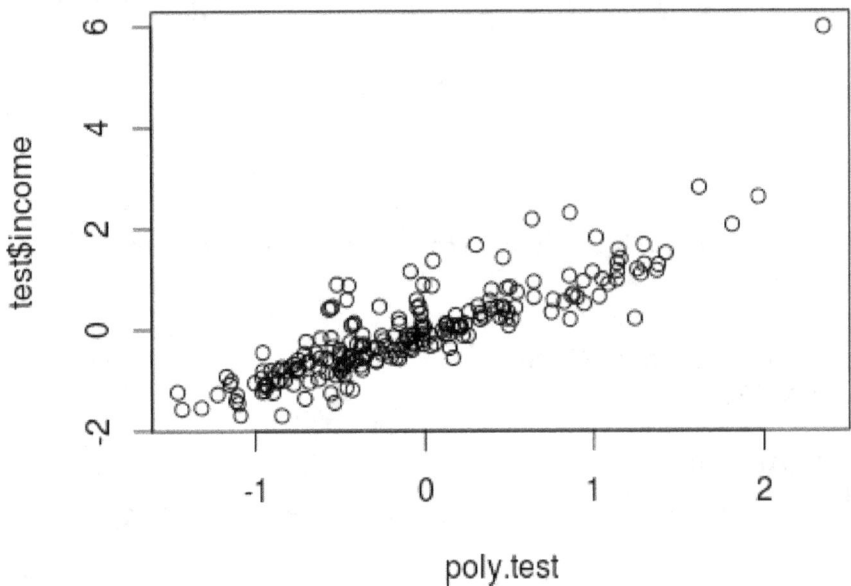

Figure 6.4: Polynomial kernel

```
poly.test.resid<-poly.test-test$income
mean(poly.test.resid^2)
## [1] 0.2453022
cor(poly.test,test$income)
## [1] 0.8563827
summary(poly.test)

   Min. 1st Qu.  Median    Mean 3rd Qu.    Max.
-1.4626 -0.5585 -0.1513 -0.0727  0.3287  2.3518
summary(test$income)
    Min.  1st Qu.   Median     Mean  3rd Qu.     Max.
-1.69649 -0.65500 -0.16247 -0.03755  0.44047  5.98180
```

The polynomial has an insignificant additional amount of error. The correlation is naturally slightly lower and the summary statistics are almost the same.

Radial Kernel

Next, we will use the radial kernel. This kernel has the parameter "gamma." This particular kernel is useful when the data is non-linear in nature. Below is the code. Figure 6.5 is the plot of the radial kernel.

```
set.seed(123)
rbf.tune<-tune.svm(income~.,data=train,kernel="radial",gamma = c(.1,.5,1,2,3,4))
best.rbf<-rbf.tune$best.model
rbf.test<-predict(best.rbf,newdata=test)
plot(rbf.test,test$income)
```

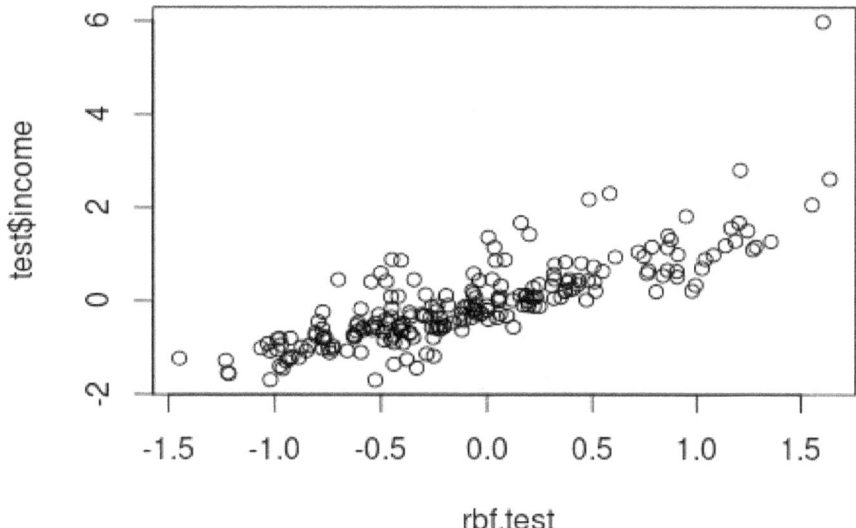

Figure 6.5: Radial kernel

```
rbf.test.resid<-rbf.test-test$income
mean(rbf.test.resid^2)
## [1] 0.3138517
cor(rbf.test,test$income)
## [1] 0.8145166
summary(rbf.test)
    Min.  1st Qu.   Median     Mean  3rd Qu.     Max.
-1.44924 -0.50971 -0.12037 -0.06814  0.31609  1.63721
summary(test$income)
    Min.  1st Qu.   Median     Mean  3rd Qu.     Max.
-1.69649 -0.65500 -0.16247 -0.03755  0.44047  5.98180
```

The radial kernel is worse than the linear and polynomial kernel. This makes sense. The results for the linear kernel were so good that the data is probably linear in nature. Since the radial kernel is best for non-linear data it should struggle with linear data. In addition, the correlation is lower (although still strong) and the difference in the summary statistics is larger.

Sigmoid Kernel

Next, we will try the sigmoid kernel. Sigmoid kernel relies on a "gamma" parameter as well as a coefficient theta. This kernel is similar to the logistic regression function. Below is the code. Figure 6.6 is the plot for the sigmoid kernel.

```
set.seed(123)
sigmoid.tune<-tune.svm(income~., data=train,kernel="sigmoid",gamma =
c(.1,.5,1,2,3,4),coef0 = c(.1,.5,1,2,3,4))
best.sigmoid<-sigmoid.tune$best.model
sigmoid.test<-predict(best.sigmoid,newdata=test)
plot(sigmoid.test,test$income)
```

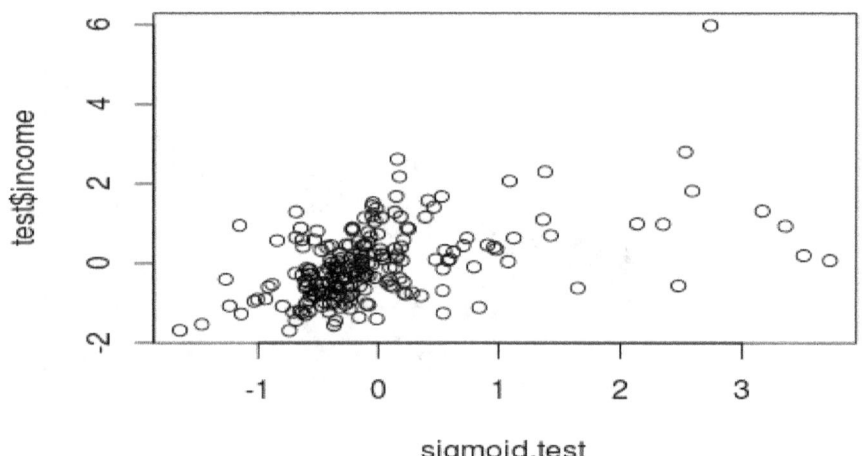

Figure 6.6: Sigmoid kernel

```
sigmoid.test.resid<-sigmoid.test-test$income
mean(sigmoid.test.resid^2)
## [1] 0.8004045
cor(sigmoid.test,test$income)
## [1] 0.4875699
summary(sigmoid.test)
    Min.  1st Qu.   Median     Mean  3rd Qu.     Max.
-1.66209 -0.45134 -0.21119 -0.02372  0.14065  3.71924
summary(test$income)
    Min.  1st Qu.   Median     Mean  3rd Qu.     Max.
-1.69649 -0.65500 -0.16247 -0.03755  0.44047  5.98180
```

The sigmoid performed much worst then the other models based on the mean squared error. You can see the mess in the plot for yourself. The correlation is also terrible compared to the other models. However, when a comparison is made based on the summary statistics the sigmoid kernel does a good job with extreme values, the mean, and the median. It would make sense that a sigmoid kernel would struggle with linear predictions as it is commonly used for classification models.

The final results for our models are as follows in figure 6.7

Place	Kernel	Mean Squared Error
1st	Linear	.21
2nd	Polynomial	.24
3rd	Radial	.31
4th	Sigmoid	.80

Figure 6.7: Kernel results

Determining which kernel to use requires trial and error, as there are no rules for selection. Using the metrics that we employed here can provide insight into what is most appropriate for numeric prediction.

Conclusion
SVM is a highly flexible machine-learning algorithm that has various forms of calculation to deal with numeric or classification predications. The example in this chapter involved numeric prediction. The power of various kernels allows you to develop several different models that can be compared that all utilize SVM. In the next chapter, we will begin our discussion on artificial neural networks.

Chapter Seven: Artificial Neural Networks

Artificial neural network (ANN) is another one of those mysterious "black box" methods in machine learning. The goal here is not to get stuck in trying to digest the complex mathematics of ANN. To try to explain it simply, this method tries to classify or predict through the use of hidden layers that each calculate different features for prediction.

Chapter Objectives
The objectives of this chapter are as follows.
- Describe artificial neural networks
- Share the pros and cons of artificial neural networks
- Prepare data for analysis
- Develop and test an artificial neural networks
- Compare the performance of different artificial neural networks

The Human Mind and the Artificial One
We will begin by looking at how real neurons work before looking at ANN. In simple terms, as this is not a biology book, neurons send and receive signals inside the human body. They receive signals through their dendrites, process information in the soma, and the send signals through their axon terminal. Below is a picture of a neuron.

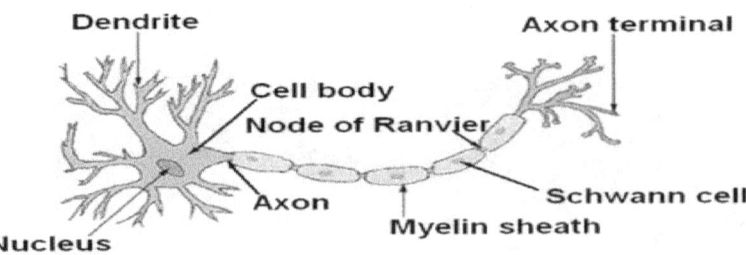

Figure 7.1: Neuron diagram

Neural Networks: Classification

An ANN works in a similar manner. The x variables are the dendrites that are providing information to the cell body where they are summed. Different dendrites or x variables can have different weights (w). Next, the summation of the x variables is passed to an activation function before moving to the output or dependent variable y, which is similar to our axon terminal in a real neuron. Below is a picture of this process.

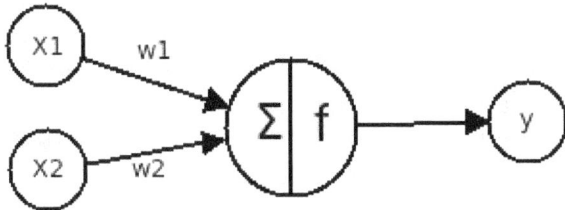

Figure 7.2: Neural Network

If you compare the two pictures they are similar yet different. ANN mimics the mind in a way that has fascinated people for over 50 years. However, the ANN community is moving away from this analogy as being unsatisfactory. Despite this it is good to be familiar with this classic explanation.

Activation Function (f)
The activation function purpose is to determine if there should be an activation. In the human body, activation takes place when the nerve cell sends the message to the next cell. This indicates that the message was strong enough to have it move forward.

The same concept applies in ANN. A signal will not be passed on unless it meets a minimum threshold. This threshold can vary depending on how the ANN is designed, the activation function chosen, and the settings the researcher provided it.

There is a lot of mathematics involved in fully understanding ANN. Some of the concepts needed to understanding the math is knowledge of Calculus, derivatives, as well as vectorization. In addition, there is an idea known as backward and forward propagation. As interesting as these topics are they are beyond the scope of this text.

ANN Design
The makeup of an ANNs can vary greatly. Some models have more than one output variable as shown below in figure 7.3.

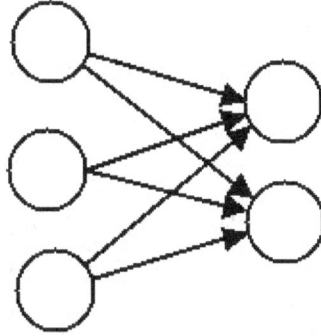

Figure 7.3 Two output ANN

All models have what are called hidden layers. These are variables that are both input and output variables. It is the researcher who determines the number of nodes in a hidden layer as well as how many hidden layers to include. Below is a visual example in figure 7.4.

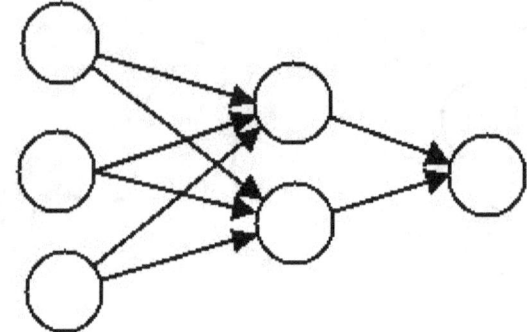

Hidden layer is the two nodes in the middle

Figure 7.4 Hidden layer artificial neural network

How many layers to develop is left to the researcher. When models become complex with several hidden layers and or outcome variables it is referred to as deep learning in the machine learning community. However, what makes an ANN deep is subjective.

Another complexity of ANN is the direction of information. Just as in the human body information can move forward and backward in an ANN using a concept known as propagation which was mentioned earlier. This provides for opportunities to model highly complex data and relationships.

ANNs are highly flexible. They can be used for both supervised and supposedly for unsupervised learning as well as for classification and numeric prediction. This flexibility is where machine learning begins to meet artificial intelligence. This type of statistical modeling is being used for many amazing things such as driverless cars and image recognition.

Pros and Cons of Artificial Neural Networks

ANN are most useful when you have a large amount of data. There is a limit to how accurate other algorithms can be when you provide massive amounts of data. However, with ANN the more data the more accurate and with enough data ANN will usually outperform any other algorithm.

However, ANN are hard to train. You need a large amount of data or you can have problems. In addition, the computational time can be excessive which makes it hard to develop models quickly. Lastly, models that are too simple often will not converge yet it is hard to know if your model is too simple until you realize it is never going to work.

An entire book could be written on ANN. For our purpose, the rest of this chapter will look at a simple application of ANN in classification.

Data Preparation

We will now go through how to create an ANN. We will be identifying people with good credit using the "GermanCredit" dataset in the "caret" package. We will make two different classification models. The first model will have a single node in the hidden layer and the second model will have a several layers in its hidden layer. In the code below, we load our libraries, and the dataset.

Neural Networks: Classification

```
library(neuralnet);library(caret)
set.seed(123)
data("GermanCredit")
```

We will not printout the number of variables using the "str" function because the list is rather long. You can examine the variable list yourself.

We now need to make some small adjustments to the data. The "class" variable is a factor variable currently but the "neuralnet" function requires numeric input only. Therefore, we will convert this variable to a dummy variable. Then we will use the "createDataPartition" function to create our training and testing sets.

```
GermanCredit$Class<-ifelse(GermanCredit$Class=="Good",1,0)
index<-createDataPartition(GermanCredit$Class,p=.7,list = F,times = 1)
trainingset<-GermanCredit[index, ]
testingset<-GermanCredit[-index, ]
```

If you looked at the data using the "str" function you will see that there were many variables in the list. Normally, when you want to include all available variables in your model when using R code you can use the "y ~.". However, the "neuralnet" function does not accept this. Since we don't want to type by hand so many variables we will need to use a shortcut. First, we will take all the names of the variables in the "GermanCredit" dataset using the "names" function.

```
gcnames<-names(GermanCredit)
```

Next, we will create a formula in which we create our equation and paste in everything except the "Class" variable. We will also include plus signs in-between each independent variable. The code is below.

```
gcformula<-as.formula(paste("Class~",paste(gcnames[!gcnames %in% "Class"],collapse = " + ")))
```

Single Hidden Layer
Model Development

We can now create our initial model. We will use the "neuralnet" function. First, we will place our formula in the parentheses. Since this model is for classification, we need to set the "err.fct" argument to "ce" which means cross-entropy. The concept of entropy should sound familiar from developing classification trees. In addition, we do not want a "linear.output" as the output is not numerical so we will set the "linear.output" argument to false. The code is as follows.

```
gcmodel<-neuralnet(gcformula,data=trainingset,err.fct = "ce",linear.output = F)
```

You can plot the model. However, because there are so many variables it is not interpretable. Figure 7.5 is a plot of our simple ANN. The code is as follows.

```
plot(gcmodel)
```

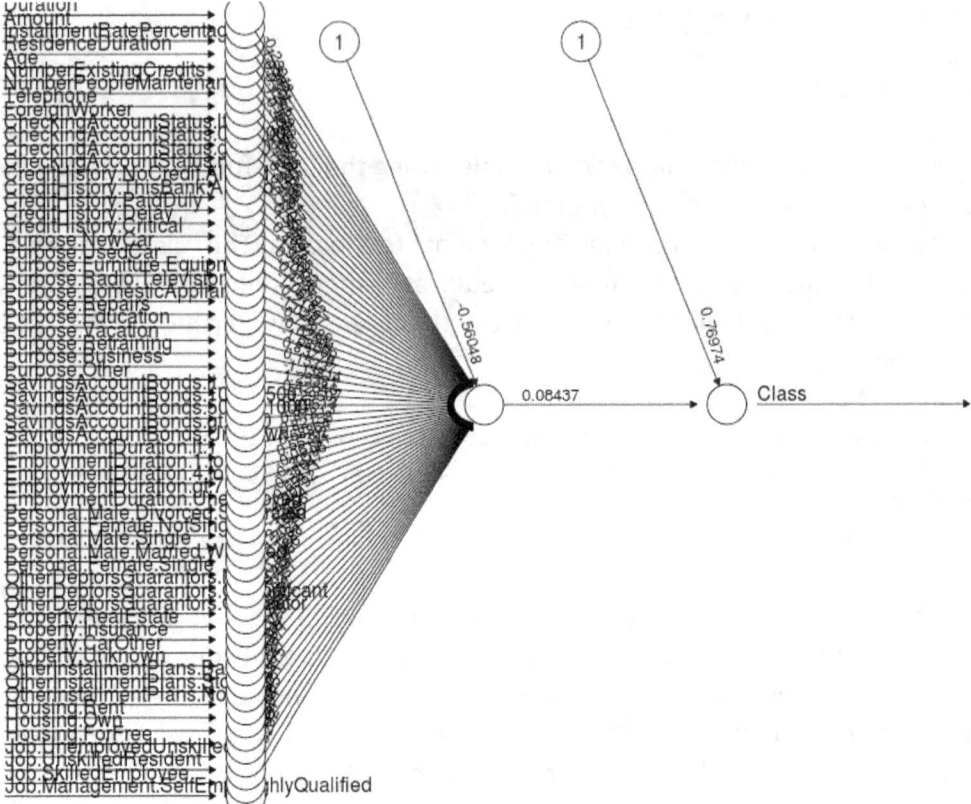

Figure 7.5: Simple ANN

Model Testing

We are now ready evaluate our model. This is done with the "compute" function. We first need to see how our model performs with the training set. Normally, this is not necessary but with the "neuralnet" function, you have to do this. We have to remove the dependent variable "class" from the "trainingset" because the dependent variable should not be in the model when predicting. For this reason, we have the -10 in brackets because "Class" is the tenth variable in the dataset when you count the columns from left to right.

```
resultsTrain<-compute(gcmodel,trainingset[,-10])
```

Now we need to get the actual results, which are stored in our "results" object under "net.results". We will store this in an object called "predTrain". However, the results are stored as probabilities so we need to convert the probabilities to either 1 or 0 with the "ifelse" function. After that, we compare the "predTrain" results to the "trainingset"

```
predTrain<-resultsTrain$net.result
predTrain<-ifelse(predTrain>=.5,1,0)
table(predTrain,trainingset$Class)
##
## predTrain    0    1
##         0  197  153
##         1   12  338
(197+338)./700
## [1] 0.7642857143
```

76% accuracy is fairly reasonable for such a simple model.

Below are the results on the testing set. The coding is the same the only difference is that we are using the "testingset"

```
resTest<-compute(gcmodel,testingset[,-10])
predTest<-resTest$net.result
predTest<-ifelse(predTest>=.5,1,0)
table(predTest,testingset[,10])
##
## predTest    0    1
##        0   71   86
##        1   20  123
(71+123)/300
## [1] 0.6466666667
```

There's a quite a drop in performance going from 76% to 65% accuracy. Perhaps adding several hidden layers will help.

Multiple Hidden Layers
Model Development
We will now turn our attention to making an ANN with a more complex hidden layer. The only thing new in the code below is the argument "hidden" we will set this as 5,4,3,2,1. This means that the first hidden layer has five neurons, the second 4 neurons, etc. Below is the code

```
set.seed(123)
gcmodelH1<-neuralnet(gcformula,data = trainingset,hidden = c(5,4,3,2,1),err.fct = 'ce',linear.output = F)
```

Figure 7.6 is the plot of the complex ANN. The plot is below.

```
plot(gcmodelH1)
```

Chapter 7

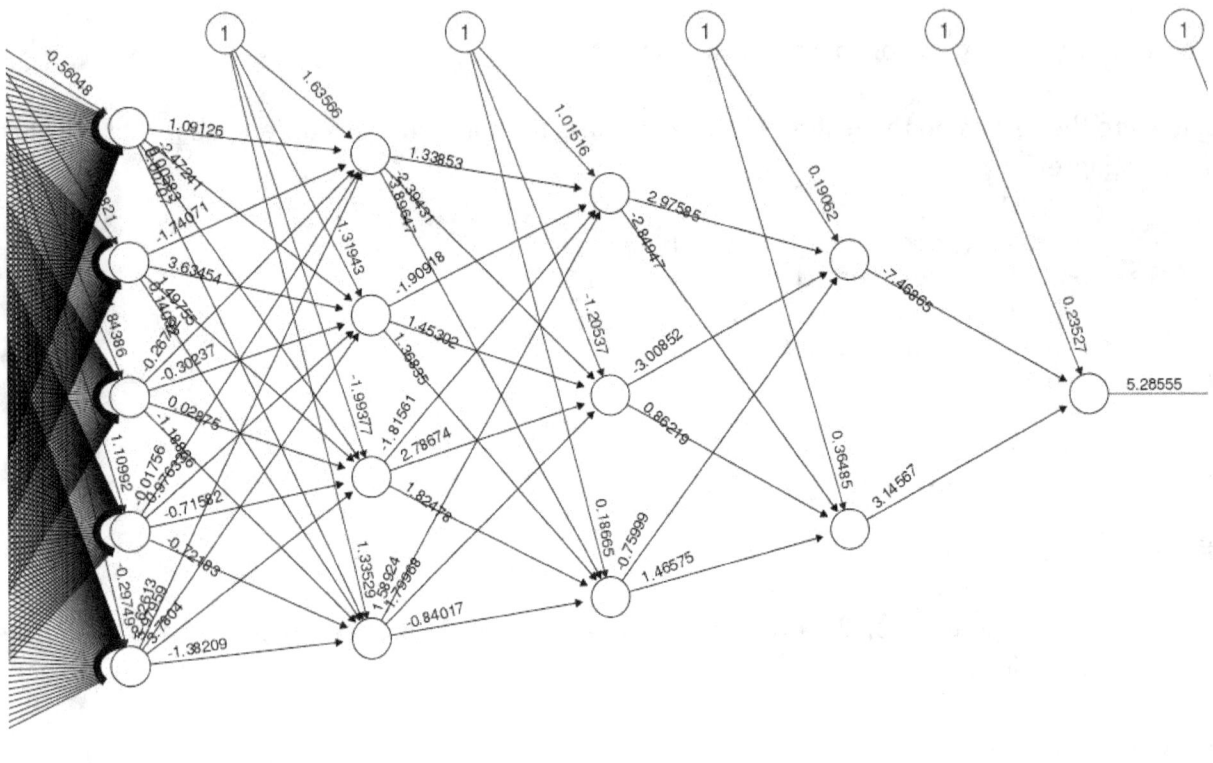

Figure 7.6: Complex ANN

You can clearly see how the model now includes the hidden layers.

Model Testing
Now we see how the model does with the training set

```
restrainH1<-compute(gcmodelH1, trainingset[,-10])
predTrainH1<-restrainH1$net.result
predTrainH1<-ifelse(predTrainH1>=.5,1,0)
table(predTrainH1,trainingset$Class)
##
## predTrainH1    0    1
##           0   47    1
##           1  162  490
(490+47)/700
## [1] 0.7671428571
```

Almost no difference in the accuracy, let's see how we will do on the testing set.

```
restestH1<-compute(gcmodelH1, testingset[,-10])
predTestH1<-restestH1$net.result
predTestH1<-ifelse(predTestH1>=.5,1,0)
table(predTestH1,testingset$Class)
##
## predTestH1    0    1
```

58

```
##           0  13   7
##           1  78 202
(202+13)/300
## [1] 0.7166666667
```

However, we did much better on the testing data at 72% accuracy. This indicates that the more complex model will be better on future datasets as the drop-off in accuracy is substantially less. How this works is somewhat mysterious and that is why it's a black box. In addition, determining the number of layers to add is also based on trial and error. Generally, the more data you have the more complex you can make your ANN.

Conclusion

ANNs is where machine learning meets artificial intelligence. The work that is being done with this algorithm is amazing. There are several variations on ANNs that were not even discussed. In general, developing models with ANNs is difficult and time-consuming and you have to avoid the temptation of just throwing all your machine learning problems at ANNs because simpler models are always preferred models. Next, we will use ANNs for numeric prediction.

Chapter Eight: Artificial Neural Networks Numeric Prediction

In this chapter, we will look at numeric prediction using artificial neural networks. We will develop both a simple network as well as one with several hidden layers. Furthermore, we will assess the model use MSE, summary statistics, as well as correlation.

Chapter Objectives
The objectives of this chapter are as follows.
- Prepare data for analysis
- Develop a single node ANN
- Develop a multiple nodes ANN
- Evaluate the models

Data Preparation
Our dataset this time is the "Wage" dataset from the "Ecdat" package. Our goal will be to predict the wage of a worker based on the other variables in the dataset. Below is some initial code.

```
library(ISLR);library(neuralnet);library(caret)
## Loading required package: lattice
## Loading required package: ggplot2
#Load data set
data("Wage")
#examine data
str(Wage)
## 'data.frame':    3000 obs. of  12 variables:
##  $ year       : int  2006 2004 2003 2003 2005 2008 2009 2008 2006 2004 ...
##  $ age        : int  18 24 45 43 50 54 44 30 41 52 ...
##  $ sex        : Factor w/ 2 levels "1. Male","2. Female": 1 1 1 1 1 1 1 1 1 1 ...
##  $ maritl     : Factor w/ 5 levels "1. Never Married",..: 1 1 2 2 4 2 2 1 1 2 ...
##  $ race       : Factor w/ 4 levels "1. White","2. Black",..: 1 1 1 3 1 1 4 3 2 1
```

```
## ...
## $ education : Factor w/ 5 levels "1. < HS Grad",..: 1 4 3 4 2 4 3 3 3 2 ...
## $ region    : Factor w/ 9 levels "1. New England",..: 2 2 2 2 2 2 2 2 2 2 ...
## $ jobclass  : Factor w/ 2 levels "1. Industrial",..: 1 2 1 2 2 2 1 2 2 2 ...
## $ health    : Factor w/ 2 levels "1. <=Good","2. >=Very Good": 1 2 1 2 1 2 2 1
2 2 ...
## $ health_ins: Factor w/ 2 levels "1. Yes","2. No": 2 2 1 1 1 1 1 1 1 1 ...
## $ logwage   : num  4.32 4.26 4.88 5.04 4.32 ...
## $ wage      : num  75 70.5 131 154.7 75 ...
```

Since the number of variables is smaller we can take more time on exploriing the data. Below is a summary of each variable

lapply(Wage, summary)

```
## $year
##    Min. 1st Qu.  Median    Mean 3rd Qu.    Max.
##    2003    2004    2006    2006    2008    2009
##
## $age
##    Min. 1st Qu.  Median    Mean 3rd Qu.    Max.
##   18.00   33.75   42.00   42.41   51.00   80.00
##
## $sex
##   1. Male 2. Female
##      3000         0
##
## $maritl
## 1. Never Married        2. Married        3. Widowed        4. Divorced
##              648              2074                19                204
##      5. Separated
##                55
##
## $race
## 1. White 2. Black 3. Asian 4. Other
##     2480      293      190       37
##
## $education
##          1. < HS Grad       2. HS Grad     3. Some College
##                   268              971                 650
##     4. College Grad 5. Advanced Degree
##                 685                426
##
## $region
##          1. New England   2. Middle Atlantic 3. East North Central
##                       0                 3000                     0
## 4. West North Central    5. South Atlantic 6. East South Central
##                       0                    0                     0
## 7. West South Central          8. Mountain            9. Pacific
##                       0                    0                     0
##
## $jobclass
```

```
##   1. Industrial 2. Information
##             1544           1456
## 
## $health
##      1. <=Good 2. >=Very Good
##             858           2142
## 
## $health_ins
## 1. Yes  2. No
##    2083    917
## 
## $logwage
##    Min. 1st Qu.  Median    Mean 3rd Qu.    Max.
##   3.000   4.447   4.653   4.654   4.857   5.763
## 
## $wage
##    Min. 1st Qu.  Median    Mean 3rd Qu.    Max.
##   20.09   85.38  104.92  111.70  128.68  318.34
```

Already you can tell that we do not need the variable "sex" as everyone is male. In addition, "logwage" is unnecessary as it's the same as "wage." Another problem we have is that we need to convert our factor variables to dummy variables. The reason being is that the "neuralnet" function does like factor variables as you know from the previous chapter. The "dummyVars" function from the "caret" package will help with this.

```
dummies<-dummyVars(wage~.,Wage,fullRank = T)
```

The next part may seem strange but bear with me. We now have an object called "dummies" which is of the class "dummyVars." However, we need a dataframe for our model because the "neuralnet" function uses dataframes. Therefore, we need to predict the "dummies" object onto our existing "Wage" dataset. After doing this we need to add to our new dataframe our dependent variable "wage" as this was lost when we created the "dummies" object. As such, the code is below.

```
Wage2<-as.data.frame(predict(dummies,newdata=Wage))
Wage2$wage<-Wage$wage
```

We also need to scale our data as this is best practice for ANN. In the previous chapter we did not scale the data as a means of keeping the amount of new information to a minimum. Now that we have a basic understanding, adding the concept of scaling should not be as overwhelming. The three variables that will be scaled are "year", "age", and "wage." It is not necessary to scale the dependent variable but we are just being consistent when we do this

```
Wage2Scaled<-as.data.frame(scale(Wage2[,c(1,2,27)]))
Wage2$year<-Wage2Scaled$year
Wage2$age<-Wage2Scaled$age
Wage2$wage<-Wage2Scaled$wage
```

We can now partition our data using the "createDataPartition" and create our training and testing set.

```
set.seed(123)
index<-createDataPartition(Wage2$wage,p=.7,list = F,times = 1)
trainingset<-Wage2[index, ]
testingset<-Wage2[-index, ]
```

Single Hidden Layer
Model Development
We are now ready to model our data. Due to the computational time of ANN we will only adjust the number of nodes in the hidden layer. We will not change the number of hidden layers in the model as it will remain at one.

```
set.seed(130)
wagemodel<-neuralnet(wage~year+age+`maritl.2. Married`+`maritl.3. Widowed`+`maritl.4. Divorced`+`maritl.5. Separated`+`race.2. Black`+`race.3. Asian`+`race.4. Other`+`education.2. HS Grad`+`education.3. Some College`+`education.4. College Grad`+`education.5. Advanced Degree`+`region.2. Middle Atlantic`+`region.3. East North Central`+`region.4. West North Central`+`region.5. South Atlantic`+`region.6. East South Central`+`region.7. West South Central`+`region.8. Mountain`+`region.9. Pacific`+`jobclass.2. Information`+`health.2. >=Very Good`+`health_ins.2. No`,data=trainingset,err.fct = "sse",linear.output = T)
```

Here is our plot in figure 8.1.

```
plot(wagemodel)
```

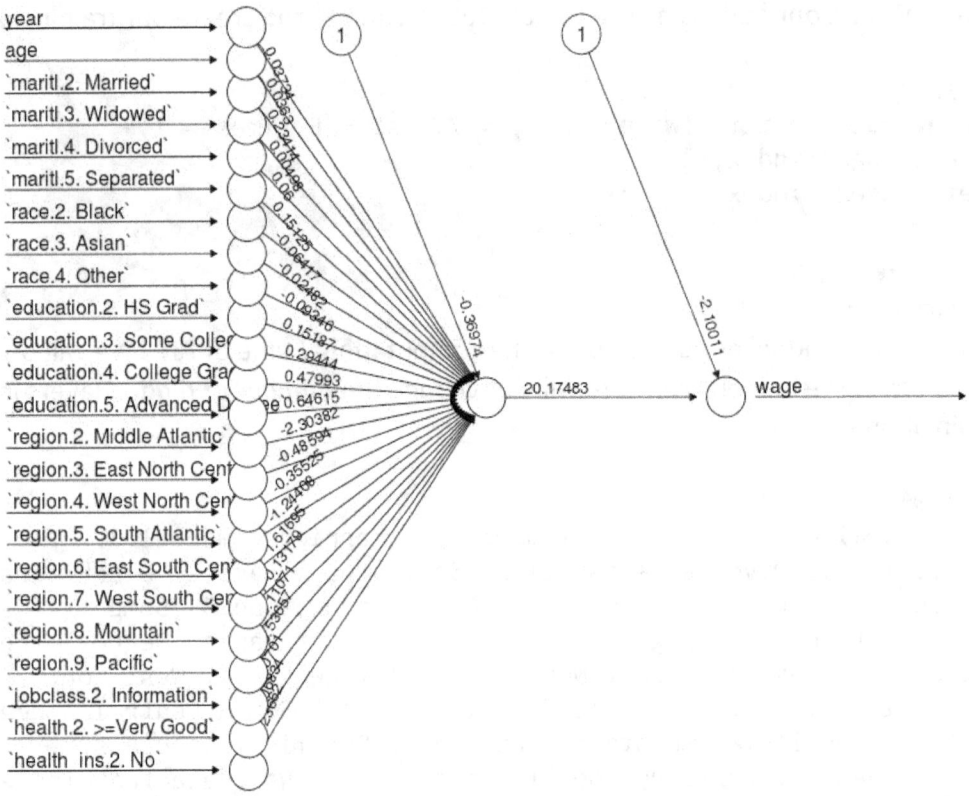

Figure 8.1: Numeric model

Model Testing

Now, we can evaluate our model first by finding how it correlates with the test set. The code below is similar to what you saw in the previous chapter. We use the "compute" function for predicting. We remove the two variables (3 = sex, 26 = logwage) that are not a part of the original. Variable 27 is the dependent variable "wage." We will also assess the correlation and the

```
evaluate_model<-compute(wagemodel, testingset[,c(-3,-26,-27)])
predicted_total<-evaluate_model$net.result
cor(predicted_total, testingset$wage)
##               [,1]
## [1,] 0.610835418
```

The correlation is strong indicating that the model works well for the test set. We will also calculate the mean absolute error. We will make a function and then run the code.

```
MAE<-function(actual, predicted){
        mean(abs(actual-predicted))
}
MAE(predicted_total,testingset$wage)
## [1] 0.5572034672
```

This number is only useful for comparison to other models.

Multiple Nodes
Model Development
We will now make a model with a hidden layer of 2 neurons.

```
set.seed(13)
wagemodel2<-neuralnet(wage~year+age+`maritl.2. Married`+`maritl.3.
Widowed`+`maritl.4. Divorced`+`maritl.5. Separated`+`race.2. Black`+`race.3.
Asian`+`race.4. Other`+`education.2. HS Grad`+`education.3. Some
College`+`education.4. College Grad`+`education.5. Advanced Degree`+`region.2.
Middle Atlantic`+`region.3. East North Central`+`region.4. West North
Central`+`region.5. South Atlantic`+`region.6. East South Central`+`region.7. West
South Central`+`region.8. Mountain`+`region.9. Pacific`+`jobclass.2.
Information`+`health.2. >=Very Good`+`health_ins.2. No`,data=trainingset,err.fct =
"sse",linear.output = T,hidden = 2)
```

The visual is below in figure 8.2.

```
plot(wagemodel2)
```

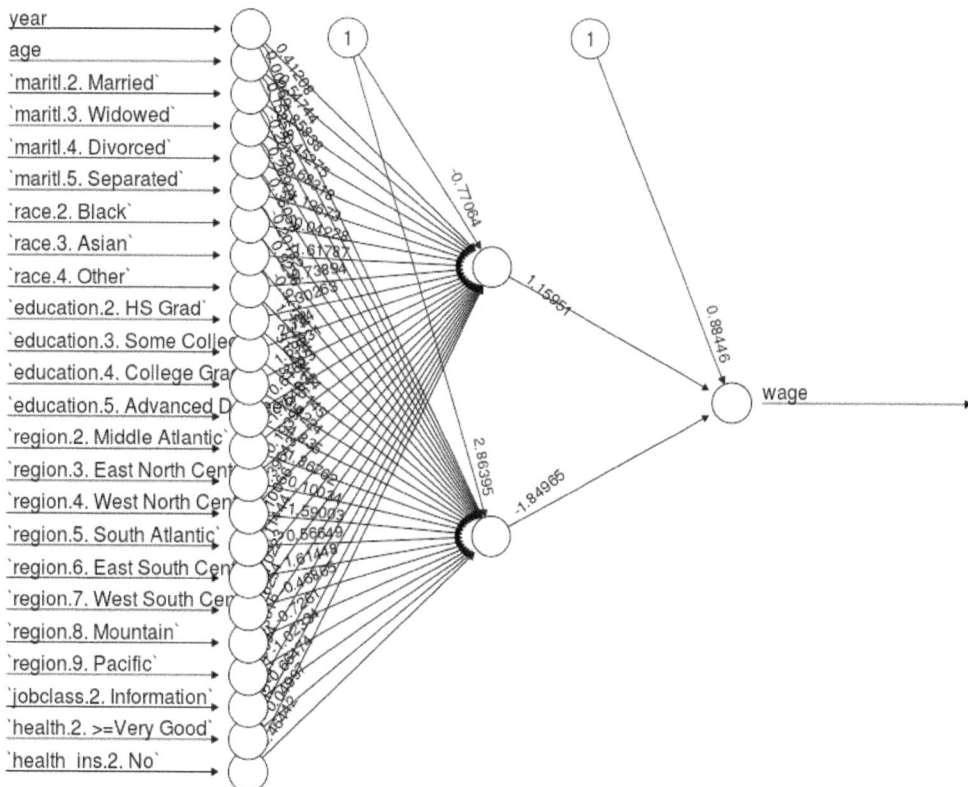

Figure 8.2: Revised model

Model Testing
We will now first check the correlation

```
evaluate_model2<-compute(wagemodel2, testingset[,c(-3,-26,-27)])
predicted_total2<-evaluate_model2$net.result
cor(predicted_total2, testingset$wage)
```

```
##                  [,1]
## [1,] 0.6027870528
```

About the same, next we will check the error

```
MAE(predicted_total2,testingset$wage)
## [1] 0.5622721841
```

There was little difference between to the two models. We could have done more but ANN takes a lot of computational power and it was a struggle to develop these two models using the current datasets. We can say confidently that our model does a fairly decent job of predicting wages. The challenge is explaining exactly how the model does this.

Conclusion

This chapter provided an example of numeric prediction with ANNs. It is important to remember that this type of modeling is used only in situation when more straightforward and simplistic approaches are unsuccessful. In the next chapter, we go deeper with deep learning H2O.

Chapter Nine: H2O Deep Learning

Deep learning is a complex machine-learning concept in which new features are created from the variables that were inputted. These new features are used for classifying labeled data are numeric prediction. This is all done mostly with artificial neural networks that are multiple layers deep and can involve regularization. In other words, deep learning is simply ANNs with several hidden layers.

The idea of "deep" is relative. For example, some will say that 3 hidden layers is deep. Another person will say that a minimum of 5 layers makes the learning deep.

If understanding is not important but you are in search of the most accurate classification possible, deep learning is a useful tool. The calculation process is nearly impossible to explain to the typical stakeholder and is best for just getting the job done. Generally, ANNs have the most complex mathematics in all of machine learning and deep learning ANNs are absolutely the most mathematically complex ideas in this discipline currently.

One of the most accessible packages for using deep learning is the "h2o" package. This package allows you to access the H2O website, which will analyze your data and send it back to you. This allows a researcher to do analytics on a much larger scale than their own computer can handle.

Chapter Objectives
The objectives of this chapter are as follows.
- Prepare data for analysis
- Develop a model for classification
- Test classification model
- Develop a numeric prediction model
- Test a numeric prediction model

The examples in this chapter are as simple as possible. There are scores of features and parameters you can adjust to use in this package. However, for the sake of simplicity we will use default features almost exclusively.

Chapter 9

Data Preparation

We will use deep learning to predict gender of the head of household in the "VietnamH" dataset from the "Ecdat" package. Some of the programming syntax is slightly different when connecting to H2O but it is understandable. Below is some initial code along with figure 9.1, which is a correlational plot of the numeric variables.

```
library(h2o);library(Ecdat);library(corrplot)
data("VietNamH")
str(VietNamH)
## 'data.frame':    5999 obs. of  11 variables:
##  $ sex     : Factor w/ 2 levels "male","female": 2 2 1 2 2 2 2 1 1 1 ...
##  $ age     : int  68 57 42 72 73 66 73 46 50 45 ...
##  $ educyr  : num  4 8 14 9 1 13 2 9 12 12 ...
##  $ farm    : Factor w/ 2 levels "yes","no": 2 2 2 2 2 2 2 2 2 2 ...
##  $ urban   : Factor w/ 2 levels "no","yes": 2 2 2 2 2 2 2 2 2 2 ...
##  $ hhsize  : int  6 6 6 6 8 7 9 4 5 4 ...
##  $ lntotal : num  10.1 10.3 10.9 10.3 10.5 ...
##  $ lnmed   : num  11.23 8.51 8.71 9.29 7.56 ...
##  $ lnrlfood: num  8.64 9.35 10.23 9.26 9.59 ...
##  $ lnexp12m: num  11.23 8.51 8.71 9.29 7.56 ...
##  $ commune : Factor w/ 194 levels "1","10","100",..: 1 1 1 1 1 1 1 1 1 1 ...
corrplot(cor(na.omit(VietNamH[,c(-1,-4,-5,-11)])),method = 'number',col= "black")
```

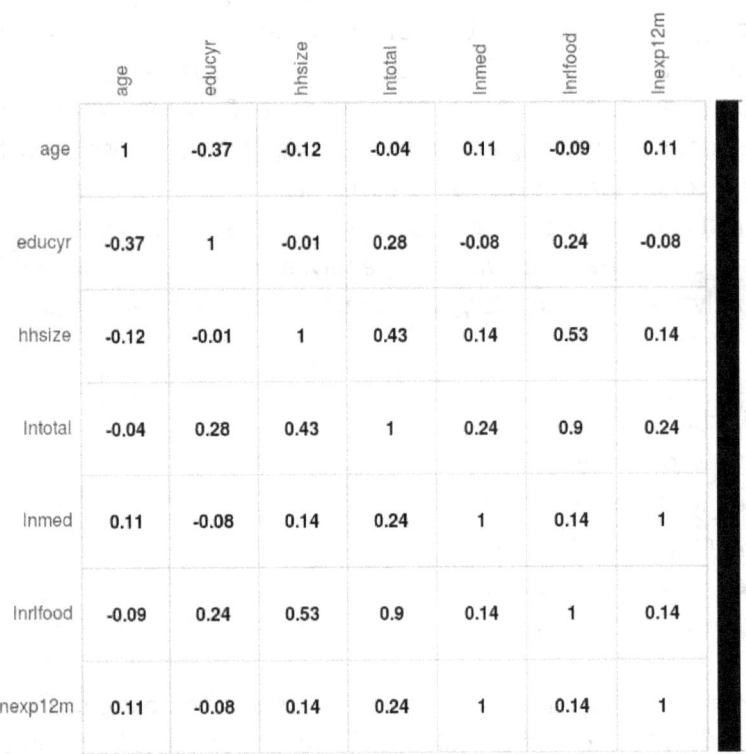

Figure 9.1: Correlational plot

We need to remove the "commune" variable "lnexp12m" and the "lntotal" variable. The "commune" variable should be removed because it does not provide much information. The "lntotal" variable should be removed because it is the total expenditures that the family spends. This is represented by other variables such as food "lnrlfood" which "lntotal" highly correlates with. The "lnexp12m" should be removed because it has a perfect correlation with "lnmed". We also need to scale the data as this was mentioned as best practice previously. Therefore, we will temporarily remove our factor variables, scale the data, and add our factor variables back. This is all done in a new object called "VietNamHscaled." Below is the code for the removal of the variables just mentioned.

```
VietNamHscaled<-as.data.frame(scale(VietNamH[,c(-1,-4,-5,-11)]))
VietNamHscaled$commune<-NULL
VietNamHscaled$lnexp12m<-NULL
VietNamHscaled$lntotal<-NULL
VietNamHscaled$farm<-VietNamH$farm
VietNamHscaled$urban<-VietNamH$urban
VietNamHscaled$sex<-VietNamH$sex
```

We now need to save our modified dataset as a csv file that we can send to h2o. The code is as follows.

```
write.csv(VietNamHscaled, file="viet.csv",row.names = F)
```

Now we can connect to H2o and start what is called an instance. The function for this is "h2O.init"

```
localH2O<-h2o.init()
##   Connection successful!
## 
## R is connected to the H2O cluster:
##      H2O cluster uptime:          5 minutes 44 seconds
##      H2O cluster version:         3.10.5.3
##      H2O cluster version age:     1 month and 28 days
##      H2O cluster name:            H2O_started_from_R_darrin_hdp730
##      H2O cluster total nodes:     1
##      H2O cluster total memory:    3.33 GB
##      H2O cluster total cores:     4
##      H2O cluster allowed cores:   4
##      H2O cluster healthy:         TRUE
##      H2O Connection ip:           localhost
##      H2O Connection port:         54321
##      H2O Connection proxy:        NA
##      H2O Internal Security:       FALSE
##      R Version:                   R version 3.4.1 (2017-06-30)
```

The output indicates that we are connected. The next step is where it really gets complicated. We need to upload our data to h2o as an h2o dataframe, which is different from a regular data frame in r. We also need to indicate the location of the csv file on our computer that needs to be converted. To complete this step we use the function "h2o.uploadFile." All of this is done in the code below.

```
viet.hex<-h2o.uploadFile(path="/home/darrin/Documents/R working
directory/blog/blog/viet.csv",destination_frame = "viet.hex")
##
  |                                                                      |   0%
  |======================================================================| 100%
```

In the code above, we create an object called "viet.hex". This object uses the "h2o.uploadFile" function to send our csv file to h2o. For the "path" argument you need to put the location of the csv file on your computer and not copy the address I used. We can check if everything worked by using the "class" function and the "str" function on "viet.hex"

class(viet.hex)

[1] "H2OFrame"

str(viet.hex)

```
## Class 'H2OFrame' <environment: 0x6ccda48>
##  - attr(*, "op")= chr "Parse"
##  - attr(*, "id")= chr "viet.hex"
##  - attr(*, "eval")= logi FALSE
##  - attr(*, "nrow")= int 5999
##  - attr(*, "ncol")= int 8
##  - attr(*, "types")=List of 8
##   ..$ : chr "real"
##   ..$ : chr "real"
##   ..$ : chr "real"
##   ..$ : chr "real"
##   ..$ : chr "real"
##   ..$ : chr "enum"
##   ..$ : chr "enum"
##   ..$ : chr "enum"
##  - attr(*, "data")='data.frame':  10 obs. of  8 variables:
##   ..$ age    : num  1.451 0.653 -0.437 1.742 1.815 ...
##   ..$ educyr : num  -0.701 0.205 1.564 0.432 -1.38 ...
##   ..$ hhsize : num  0.638 0.638 0.638 0.638 1.662 ...
##   ..$ lnmed  : num  2.162 1.173 1.249 1.458 0.829 ...
##   ..$ lnrlfood: num  -0.0749 1.2411 2.8814 1.0883 1.7014 ...
##   ..$ farm   : Factor w/ 2 levels "no","yes": 1 1 1 1 1 1 1 1 1 1
##   ..$ urban  : Factor w/ 2 levels "no","yes": 2 2 2 2 2 2 2 2 2 2
##   ..$ sex    : Factor w/ 2 levels "female","male": 1 1 2 1 1 1 1 2 2 2
```

The "summary" function also provides insight into the data.

summary(viet.hex)

```
## Warning in summary.H2OFrame(viet.hex): Approximated quantiles
## computed! If you are interested in exact quantiles, please pass the
## `exact_quantiles=TRUE` parameter.
```

```
##    age                  educyr              hhsize
##  Min.   :-2.325e+00   Min.   :-1.606e+00   Min.   :-1.920e+00
##  1st Qu.:-8.045e-01   1st Qu.:-7.048e-01   1st Qu.:-3.911e-01
##  Median :-1.505e-01   Median :-2.229e-02   Median : 1.247e-01
##  Mean   :-1.630e-16   Mean   :-8.313e-17   Mean   : 7.324e-16
##  3rd Qu.: 7.216e-01   3rd Qu.: 6.552e-01   3rd Qu.: 6.313e-01
##  Max.   : 3.412e+00   Max.   : 3.375e+00   Max.   : 7.290e+00
##    lnmed               lnrlfood             farm       urban
##  Min.   :-1.908e+00   Min.   :-4.328e+00   yes:3438   no :4269
##  1st Qu.:-3.984e-01   1st Qu.:-5.722e-01   no :2561   yes:1730
##  Median : 2.511e-01   Median : 1.785e-02
##  Mean   :-9.308e-16   Mean   :-9.709e-16
##  3rd Qu.: 6.900e-01   3rd Qu.: 5.985e-01
##  Max.   : 2.571e+00   Max.   : 5.038e+00
##     sex
##  male  :4375
##  female:1624
##
##
##
##
```

We now need to create our train and test sets. We need to use slightly different syntax to do this with h2o. The code below is how it is done to create a 70/30 split in the data.

```
rand<-h2o.runif(viet.hex,seed = 123)
train<-viet.hex[rand<=.7,]
train<-h2o.assign(train, key = "train")
test<-viet.hex[rand>.7,]
test<-h2o.assign(test, key = "test")
```

Here is what we did
1. We create an object called "rand" that created random numbers for our "viet.hex" dataset.
2. All values less than .7 were assigned to the train variable
3. The train variable was give the key name "train" in order to use it in the h2o framework
4. All values greater than .7 were assigned to test and "test" was given as the key name

You can check the proportions of the train and test sets using the "h2o.table" function.

```
h2o.table(train$sex)
##       sex Count
## 1 female  1146
## 2   male  3058
##
## [2 rows x 2 columns]
h2o.table(test$sex)
##       sex Count
## 1 female   478
## 2   male  1317
##
## [2 rows x 2 columns]
```

Classification Model
Model Development and Testing

We can now create are model called "vietdlmodel" and call it in the r environment.

vietdlmodel<-**h2o.deeplearning**(x=1:7,y=8,training_frame = train,validation_frame = test,seed=123,variable_importances = T)

```
##
  |===================================================================| 100%
vietdlmodel

## Model Details:
## ==============
##
## H2OBinomialModel: deeplearning
## Model ID:  DeepLearning_model_R_1505350138261_12
## Status of Neuron Layers: predicting sex, 2-class classification, bernoulli distr
ibution, CrossEntropy loss, 43,002 weights/biases, 512.9 KB, 42,040 training sample
s, mini-batch size 1
##   layer units      type dropout       l1       l2 mean_rate rate_rms
## 1     1    11     Input  0.00 %
## 2     2   200  Rectifier  0.00 % 0.000000 0.000000  0.187295 0.391064
## 3     3   200  Rectifier  0.00 % 0.000000 0.000000  0.303972 0.366521
## 4     4     2   Softmax         0.000000 0.000000  0.005715 0.004347
##   momentum mean_weight weight_rms mean_bias bias_rms
## 1
## 2 0.000000   -0.042509   0.101447  0.220270 0.076391
## 3 0.000000   -0.026378   0.075372  0.703335 0.185027
## 4 0.000000   -0.029633   0.337783  0.000062 0.023520
##
##
## H2OBinomialMetrics: deeplearning
## ** Reported on training data. **
## ** Metrics reported on full training frame **
##
## MSE:  0.1450856
## RMSE:  0.380901
## LogLoss:  0.4523682
## Mean Per-Class Error:  0.3362085
## AUC:  0.8097643
## Gini:  0.6195286
##
## Confusion Matrix (vertical: actual; across: predicted) for F1-optimal threshold:
##        female male    Error      Rate
## female    429  717 0.625654  =717/1146
## male      143 2915 0.046763  =143/3058
## Totals    572 3632 0.204567  =860/4204
##
## Maximum Metrics: Maximum metrics at their respective thresholds
##                       metric threshold    value idx
## 1                     max f1  0.494462 0.871450 271
```

```
## 2                        max f2  0.303325 0.936170 336
## 3                 max f0point5  0.722625 0.851488 167
## 4                  max accuracy 0.580135 0.801618 235
## 5                 max precision 0.982799 1.000000   0
## 6                    max recall 0.077476 1.000000 396
## 7               max specificity 0.982799 1.000000   0
## 8              max absolute_mcc 0.580135 0.461692 235
## 9       max min_per_class_accuracy 0.755399 0.734729 150
## 10     max mean_per_class_accuracy 0.759776 0.737981 147
##
## Gains/Lift Table: Extract with `h2o.gainsLift(<model>, <data>)` or `h2o.gainsLif
t(<model>, valid=<T/F>, xval=<T/F>)`
## H2OBinomialMetrics: deeplearning
## ** Reported on validation data. **
## ** Metrics reported on full validation frame **
##
## MSE:  0.1510359
## RMSE:  0.3886334
## LogLoss:  0.4648798
## Mean Per-Class Error:  0.379009
## AUC:  0.7937798
## Gini:  0.5875595
##
## Confusion Matrix (vertical: actual; across: predicted) for F1-optimal threshold:
##         female male    Error       Rate
## female     132  346 0.723849   =346/478
## male        45 1272 0.034169    =45/1317
## Totals     177 1618 0.217827   =391/1795
##
## Maximum Metrics: Maximum metrics at their respective thresholds
##                        metric threshold    value idx
## 1                      max f1  0.424213 0.866780 302
## 2                      max f2  0.208051 0.935452 373
## 3               max f0point5  0.784682 0.840939 136
## 4                max accuracy 0.523840 0.785515 264
## 5               max precision 0.982497 1.000000   0
## 6                  max recall 0.120171 1.000000 391
## 7             max specificity 0.982497 1.000000   0
## 8            max absolute_mcc 0.784682 0.406078 136
## 9     max min_per_class_accuracy 0.760672 0.717573 151
## 10   max mean_per_class_accuracy 0.811119 0.728654 120
##
## Gains/Lift Table: Extract with `h2o.gainsLift(<model>, <data>)` or `h2o.gainsLif
t(<model>, valid=<T/F>, xval=<T/F>)`
```

Here is what the code above means.
1. We created an object called "vietdlmodel"
2. We used the "h2o.deeplearning" function.
3. x = 1:7 are all the independent variables in the dataframe and y=8 is the dependent variable "sex"

4. We set the training and testing frame to "train" and "test" and set the seed.
5. Finally, we indicated that we want to know the variable importance.

In the output above, there are results for the training data and the testing data (called the validation set). In addition, all the metrics are calculated for us. For the training data, the MSE was .14 and the error rate was about 20%. For the test data, the MSE was .15 and the error rate was 21%. Lastly, at the top of the output it indicates that this was a four layer ANN.

The output provides confusion matrices for both datasets as well. There are also other metrics called "Maximum Metrics" that use a combination of true/false positives and true/false negatives to compute the various metrics here.

Since the h2o output is so thorough we do not need to create a table ourselves to compute the accuracy. We can also look at the importance of each variable. Below we can see which variable were most useful.

vietdlmodel@model$variable_importances

```
## Variable Importances:
##              variable relative_importance scaled_importance percentage
## 1            urban.no            1.000000          1.000000   0.192573
## 2           urban.yes            0.860112          0.860112   0.165634
## 3             farm.no            0.819853          0.819853   0.157882
## 4            farm.yes            0.710458          0.710458   0.136815
## 5                 age            0.478829          0.478829   0.092210
## 6              hhsize            0.372760          0.372760   0.071784
## 7             lnrlfood           0.366639          0.366639   0.070605
## 8              educyr            0.343933          0.343933   0.066232
## 9               lnmed            0.240246          0.240246   0.046265
## 10    farm.missing(NA)            0.000000          0.000000   0.000000
## 11   urban.missing(NA)            0.000000          0.000000   0.000000
```

The numbers speak for themselves. "Urban" and "farm" are both the most important variables for predicting sex. Below is the code for placing the predicted results into a dataframe. If you wanted to send the results to a competition such as kaggle.

vietdlPredict<-h2o.predict(vietdlmodel,newdata = test)
```
##
  |
  |                                                                 |   0%
  |
  |=================================================================| 100%
```
vietdlPredict

```
##   predict    female      male
## 1    male 0.1583503 0.8416497
## 2    male 0.2335769 0.7664231
## 3    male 0.2503513 0.7496487
## 4    male 0.3059966 0.6940034
## 5    male 0.1403371 0.8596629
## 6    male 0.2126763 0.7873237
```

```
## 
## [1795 rows x 3 columns]
```

```r
vietdlPred<-as.data.frame(vietdlPredict)
head(vietdlPred)
```

```
##   predict    female      male
## 1    male 0.1583503 0.8416497
## 2    male 0.2335769 0.7664231
## 3    male 0.2503513 0.7496487
## 4    male 0.3059966 0.6940034
## 5    male 0.1403371 0.8596629
## 6    male 0.2126763 0.7873237
```

Numeric Prediction

Model Development and Testing

The code below is how to conduct numeric prediction using H2O. We will use the same data set "VietNamH" but now we will predict total household expenditures "lntotal."

```r
library(h2o);library(Ecdat)
data("VietNamH")
...
```

You know from the correlational plot from earlier that we need to remove the "commune, and variable and the "lnexp12m" variable. The "commune" variable should be removed because it doesn't provide much information. The "lnexp12m" variable should be removed because it is the total expenditures that the family spends which highly correlates "lntotal". In addition, the "lnexp12m" should be removed because it has a perfect correlation with "lnmed".

Therefore, we will scale our data again, remove the useless variables, and add in again the factor variables. Below is the code

```r
VietNamHscaled<-as.data.frame(scale(VietNamH[,c(-1,-4,-5,-11)]))
VietNamHscaled$commune<-NULL
VietNamHscaled$lnexp12m<-NULL
VietNamHscaled$farm<-VietNamH$farm
VietNamHscaled$urban<-VietNamH$urban
VietNamHscaled$sex<-VietNamH$sex
```

We now need to save our modified dataset as a csv file that we can send to h2o. The code is as follows.

```r
write.csv(VietNamH, file="viet.csv",row.names = F)
```
Now we can connect to H2o as before. This should still be mostly review

```r
localH2O<-h2o.init()
##   Connection successful!
## 
## R is connected to the H2O cluster:
##     H2O cluster uptime:         1 hours 38 minutes
##     H2O cluster version:        3.10.5.3
```

```
##         H2O cluster version age:    1 month and 28 days
##         H2O cluster name:           H2O_started_from_R_darrin_hdp730
##         H2O cluster total nodes:    1
##         H2O cluster total memory:   3.33 GB
##         H2O cluster total cores:    4
##         H2O cluster allowed cores:  4
##         H2O cluster healthy:        TRUE
##         H2O Connection ip:          localhost
##         H2O Connection port:        54321
##         H2O Connection proxy:       NA
##         H2O Internal Security:      FALSE
##         R Version:                  R version 3.4.1 (2017-06-30)
```

The output indicates that we are connected. Now, we need to upload our data to h2o as an h2o dataframe and indicate the location of the csv file on our computer that needs to be converted.

```
viet.hexreg<-h2o.uploadFile(path="/home/darrin/Documents/R working
directory/blog/blog/viet.csv",destination_frame = "viet.hex")
##
  |
  |                                                              |   0%
  |
  |==============================================================| 100%
```

We now need to create our train and test sets. The code below is how it is done to create a 70/30 split in the data.

```
rand<-h2o.runif(viet.hexreg,seed = 123)
train<-viet.hexreg[rand<=.7,]
train<-h2o.assign(train, key = "train")
test<-viet.hexreg[rand>.7,]
test<-h2o.assign(test, key = "test")
```

We can now create are model

```
vietdlmodelreg<-h2o.deeplearning(x=c(1:3,5:9),y=4,training_frame = train,validation_frame = test,seed=123,variable_importances = T)

##
  |
  |                                                              |   0%
  |
  |==============================================================| 100%
vietdlmodelreg

## Model Details:
## ==============
##
## H2ORegressionModel: deeplearning
## Model ID:  DeepLearning_model_R_1505350138261_23
## Status of Neuron Layers: predicting lntotal, regression, gaussian distribution,
  Quadratic loss, 43,401 weights/biases, 517.7 KB, 42,040 training samples, mini-batc
```

```
h size 1
##   layer units    type     dropout       l1       l2 mean_rate rate_rms
## 1     1    14    Input      0.00 %
## 2     2   200 Rectifier    0.00 % 0.000000 0.000000  0.228296 0.407045
## 3     3   200 Rectifier    0.00 % 0.000000 0.000000  0.443775 0.333019
## 4     4     1   Linear             0.000000 0.000000  0.005341 0.003965
##   momentum mean_weight weight_rms mean_bias bias_rms
## 1
## 2 0.000000   -0.030329   0.103272  0.304806 0.092961
## 3 0.000000   -0.022700   0.074760  0.848458 0.061910
## 4 0.000000   -0.002330   0.037329  0.026546 0.000000
##
##
## H2ORegressionMetrics: deeplearning
## ** Reported on training data. **
## ** Metrics reported on full training frame **
##
## MSE:  0.1371156
## RMSE:  0.3702912
## MAE:  0.270345
## RMSLE:  NaN
## Mean Residual Deviance :  0.1371156
##
##
## H2ORegressionMetrics: deeplearning
## ** Reported on validation data. **
## ** Metrics reported on full validation frame **
##
## MSE:  0.1385855
## RMSE:  0.3722707
## MAE:  0.2762324
## RMSLE:  NaN
## Mean Residual Deviance :  0.1385855
```

Here is what the code above means.
1. We created an object called "vietdlmodelreg"
2. We used the "h2o.deeplearning" function.
3. x = 1:3 and 5:9 are all the independent variables in the dataframe and y=4 is the dependent variable "lntotal"
4. We set the training and testing frame to "train" and "test" and set the seed.
5. Finally, we indicated that we want to know the variable importance.

The MSE, RMSE, and other error measures are reported for both the training and testing data in the output above. The MSE are essentially the same for both datasets (.14 & .14). The model does equally well on both datasets, indicating that the model is generalizable. You can also assess variable importance as show in the code below.

```
vietdlmodelreg@model$variable_importances

## Variable Importances:
##               variable relative_importance scaled_importance percentage
## 1             urban.no            1.000000          1.000000   0.116278
## 2             farm.yes            0.996003          0.996003   0.115813
## 3             sex.male            0.979784          0.979784   0.113927
## 4            urban.yes            0.903077          0.903077   0.105008
## 5             lnrlfood            0.885425          0.885425   0.102955
## 6              farm.no            0.842095          0.842095   0.097917
## 7           sex.female            0.762324          0.762324   0.088641
## 8                lnmed            0.688835          0.688835   0.080096
## 9               hhsize            0.528639          0.528639   0.061469
## 10                 age            0.509149          0.509149   0.059203
## 11              educyr            0.504784          0.504784   0.058695
## 12    farm.missing(NA)            0.000000          0.000000   0.000000
## 13   urban.missing(NA)            0.000000          0.000000   0.000000
## 14     sex.missing(NA)            0.000000          0.000000   0.000000
```

The numbers speak for themselves. "urban" and "sex" are both the most important variables for predicting total household expenditures. Below is the code for taking placing the predicted results into a dataframe. This is useful if you wanted to send the results to a competition such as kaggle.

```
vietdlPredictreg<-h2o.predict(vietdlmodelreg,newdata = test)
##
  |                                                                      |   0%
  |======================================================================| 100%
vietdlPredictreg

##     predict
## 1 1.5192266
## 2 2.1541992
## 3 0.7488361
## 4 1.1245858
## 5 2.1243108
## 6 2.3414026
##
## [1795 rows x 1 column]

vietdlPredreg<-as.data.frame(vietdlPredictreg)
head(vietdlPredreg)

##     predict
## 1 1.5192266
## 2 2.1541992
## 3 0.7488361
## 4 1.1245858
```

```
## 5 2.1243108
## 6 2.3414026
```

Conclusion

H2o can be a fun experience if you are familiar with how it works. Here we saw how this package can be used both for classification and numeric prediction. The flexibility of neural networks makes them highly useful in so many ways beyond traditional machine learning. In the next chapter, we take a closer look at evaluating model performance.

Chapter Ten: Evaluating Model Performance

We will now look at assessing model performance much more closely. Many of the topics covered in this chapter are concepts we have been using throughout this book without explaining what they are. The reason for this is that I wanted to hit the ground running by using different techniques rather than spend several chapters in the beginning talking about all of the various machine-learning tools theoretically. Experience has shown me that people learn better from doing and then reviewing what they did rather than listening and then doing what they heard.

Specifically we will look at how to assess the performance of a classification model such as a model that uses decision trees, classification rules, and ANNs. In addition, there will be a brief review of assessing numeric prediction models as well. Our primary objectives in this chapter are as follows

Chapter Objectives
- Explain the details of a confusion matrix
- Assess classification models with output from a confusion matrix
- Review numeric prediction metrics

Confusion Matrix

We will begin by looking at a confusion matrix. In order to create a confusion matrix we need to create a model that classifies something. As such, we will use the "Wage" dataset in the "Ecdat" package. We want to predict a person's jobclass based on the other variables (wage) using a decision tree. We will run the code and then it will be followed with an explanation of a confusion matrix. Below is the code for the initial model.

```
library(ISLR);library(caret);library(rpart);library(e1071);library(rattle);library(rpart);library(rpart.plot)
data("Wage")
inTrain<-createDataPartition(y=Wage$jobclass,p=0.7, list=FALSE)
```

```
trainingset <- Wage[inTrain, ]
testingset <- Wage[-inTrain, ]
set.seed(1)
TreeModel<-train(jobclass~health_ins+age+wage+education+maritl+race+health,
method='rpart', data=trainingset)
```

Here is what our model looks like in figure 10.1.

fancyRpartPlot(TreeModel$finalModel)

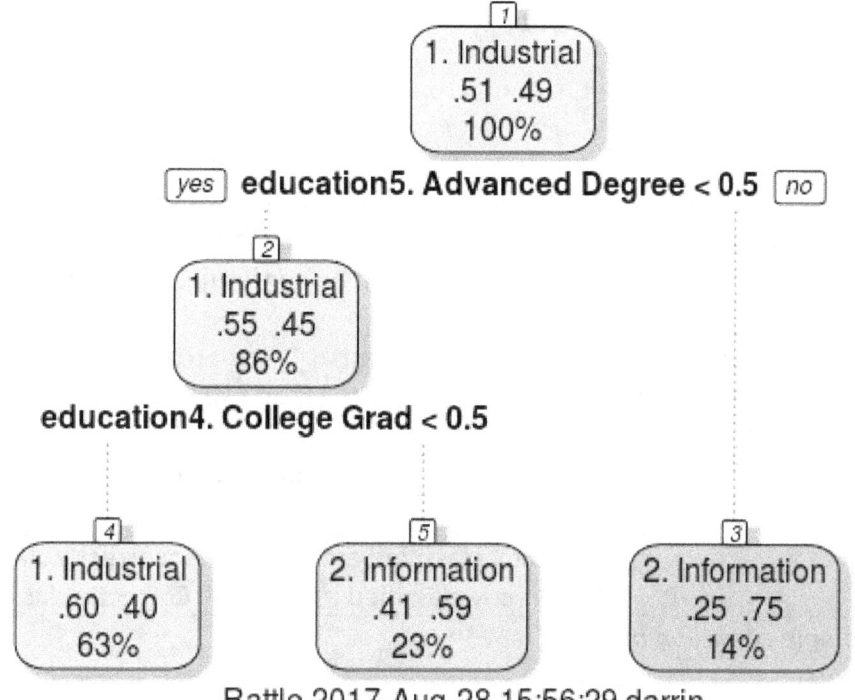

Figure 10.1: Tree model

You can predict individual values in the data set by using the 'predict' function with the test data. After predicting, we will use the "confusionMatrix" function from the "caret" package to learn how well our model did. The code is below.

```
testmodel<-predict(TreeModel, newdata = testingset)
confusionMatrix(testmodel, testingset$jobclass)
## Confusion Matrix and Statistics
## 
##                   Reference
## Prediction        1. Industrial 2. Information
##    1. Industrial            376            190
##    2. Information           106            226
## 
##                Accuracy : 0.6704
##                  95% CI : (0.6386, 0.7011)
##     No Information Rate : 0.5367
##     P-Value [Acc > NIR] : 2.755e-16
```

```
##
##                   Kappa : 0.3279
##  Mcnemar's Test P-Value : 1.405e-06
##
##             Sensitivity : 0.7801
##             Specificity : 0.5433
##          Pos Pred Value : 0.6643
##          Neg Pred Value : 0.6807
##              Prevalence : 0.5367
##          Detection Rate : 0.4187
##    Detection Prevalence : 0.6303
##       Balanced Accuracy : 0.6617
##
##         'Positive' Class : 1. Industrial
##
```

You've seen the output, at least partially, in the table above but we have not really explained it explicitly. A confusion matrix is a table that is used to organize the predictions made during an analysis of data. Without making a joke, confusion matrices can be confusing, especially for those who are new to research.

The example above is a two-class confusion matrix. This matrix compares the actual class of an example with the predicted class of the model. Table 10.2 is an example of a confusion matrix.

		Reference Class	
Predicted Class		A	B
	A	1. Correctly classified as A	4. Incorrectly classified as A
	B	3. Incorrectly classified as B	2. Correctly classified as B

Figure 10.2: Example confusion matrix

As you can see, sometimes things are classified correctly and sometimes they are not. If you look closely at figure 10.2, each column was given a number. Below is what they represent.

1. Correctly classified as A-This means that the example was a part of the A category and the model predicted it as such. This is also known as true positives
2. Correctly classified as B-This means that the example was a part of the B category and the model predicted it as such. Also known as true negatives
3. Incorrectly classified as B-This means that the example was a part of the A category but the model predicted it to be a part of the B group. Another term for this is false negatives
4. Incorrectly classified as A-This means that the example was a part of the B category but the model predicted it to be a part of the A group. This is also known as false positives.

Figure 10.3 is the results from our own R output using our classification tree. Let's see if we can determine the different classifications that took place

Evaluating Model Performance

	Reference Class	
Predicted Class	Industrial	Information
Industrial	376	190
Information	106	226

Figure 10.3: Model confusion matrix

Here is a breakdown of the results
- We have 376 true positives (Industrial correctly classified as Industrial)
- We have 226 true negatives (Information correctly classified as Information)
- We have 190 false positives (Industrial incorrectly classified as Information)
- We have 106 false negatives (Information incorrectly classified as Industrial)

One of the simplest ways to evaluate a model is to examine the confusion matrix

Assessing Models with Confusion Matrices Outputs

So what does the confusion matrix information tell us? We will look at several commonly used measures, specifically...
- Accuracy
- kappa
- error
- sensitivity
- specificity

We will look at each of these using the output from our decision tree model.

Accuracy

Accuracy is probably the easiest statistic to understand. Accuracy is the total number of items correctly classified divided by the total number of items below is the equation

$$\text{accuracy} = TP + TN / TP + TN + FP + FN$$

TP = true positive, TN = true negative, FP = false positive, FN = false negative

In our model, the accuracy was 67%

Accuracy can range in value from 0-1 with one representing 100% accuracy. Normally, you do not want perfect accuracy as this is an indication of overfitting and your model will probably not do well with other data.

Kappa

The kappa statistic is a measurement of accuracy of a model while taking into account chance. The closer the value is to 1 the better. The kappa for our model was .32 This is not too impressive.

Error

Error is the opposite of accuracy and represents the percentage of examples that are incorrectly classified. This information is not in our matrices above but its equation is as follows.

$$error = FP + FN/ TP + TN + FP + FN$$

Another way to see this equation is error = 1 – accuracy. The lower the error the better the model is in general. However, if error is 0 it indicates overfitting. Keep in mind that error is the inverse of accuracy. As one increases the other decreases.

Sensitivity

Sensitivity is the proportion of true positives that were correctly classified. The formula is as follows

$$sensitivity = TP/ TP + FN$$

This may sound confusing but high sensitivity is useful for assessing a negative result. In other words, if I am testing people for a disease and my model has a high sensitivity. This means that the model is useful telling me a person does not have a disease. For our model, the sensitivity was .78. What this means is that our model was somewhat accurate in identifying people who were really of the job class "Industrial."

Specificity

Specificity measures the proportion of negative examples that were correctly classified. The formula is below

$$specificity = TN/ TN + FP$$

Returning to the disease example, a high specificity is a good measure for determining if someone has a disease if he or she test positive for it. The specificity for our model was .54. Remember that no test is foolproof and there are always false positives and negatives happening. The role of the researcher is to maximize the sensitivity or specificity based on the purpose of the model.

There are other metrics in the confusion matrix that we are not going to discuss such as Mcnemar's test, which is useful in a medical context However, there are also several metrics that the confusion matrix does not calculate, and these are listed below.

Precision

For example, Precision is the proportion of examples that are really positive. The formula is as follows

$$precision = TP/ TP + FP$$

The more precise a model is the more trustworthy it is. In other words, high precision indicates that the results are relevant. For the our model it was .78

Recall

Recall is a measure of the completeness of the results of a model. It is calculated as follows

recall = TP/ TP + FN

This formula is the same as the formula for sensitivity. The difference is in the interpretation. High recall means that the results have a breadth to them such as in search engine results. For our model the recall was .66

F-Measure

The f-measure uses recall and precision to develop another way to assess a model. The formula is below

F-Measure = 2 * TP / 2 * TP + FP + FN

The f-measure can range from 0 – 1 and is useful for comparing several potential models using one convenient number. For our model the f-measure is .55.

Numeric Prediction

Improving numeric prediction was demonstrated in every chapter of this book that addressed numeric prediction. Therefore, we will not go through additional examples here. Instead, below is a list of some of the most common ways to assess the performance of a numeric prediction model.

- the error (RMSE, MSE, MAE)
- correlation between predicted and actual results
- Model specific metric (r^2, AIC)
- Summary statistics

Model performance is always relative. In other words, good or bad is determined by an arbitrary standard that you set. As one statistician once said, "all models are wrong but some are useful." It is your responsibility to define what "useful" is.

Conclusion

This chapter provided an in-depth explanation of how to evaluate model performance. Evaluation is a subjective and comparative experience. When judging a models quality it is always important to remember what the purpose and goals of the final model are. In the final chapter, we will look at how to improve model performance.

Chapter Eleven: Improving Model Performance

There are times when it is necessary to fine-tune a model to meet challenges in the data. The process of fine tuning involves the manipulation of parameter(s). By manipulating this value(s) you can change the predictions of the algorithm. When this is needed, the easiest thing to do is to employ the "caret" package for hyperparameter tuning. In this chapter, we will take a look at the detailed oriented process of model-tuning.

Chapter Objectives
- Explain the role of parameters in model-tuning
- Allow "caret" to automatically tune a model
- Tune a model manually

Hyperparameters
A hyperparameter in machine learning is a value that is set before the algorithm begins the learning process. Each model or algorithm has a different hyperparameter(s) that can be tuned. For example, for support vector machines the hyperparameter that you can tune is "C" which stands for cost function, which is the penalty for misclassification or numeric prediction error. For random forest, the hyperparameter that can be tuned is "mtry" which is the number of random variables selected for each split in a tree.

Regular decision trees have three potential hyperparameters to tune and they are "model", "trials", and "winnow." Regression trees have "cp" which means "complexity parameter and influences the size of trees. Modal tress have "pruned" "smoothed", and "rules." Lastly, classification rules have the "NumOpt" hyperparameter.

There are two ways to adjust hyperparameters when using the "caret" package
- Let "caret" do it
- You do it

Automatically Tuned Model

If "caret" does the work, it will pick three values for each hyperparameter to estimate separate models. This leads to an equation of 3^p. Therefore, if we have two hyperparameters we would make nine models (3^2 = 9). The "caret" package will tell you which combination of hyperparameters is best and you will then set the hyperparameters in the model to the values. How the values of the hyperparameters are set depends on the algorithm you are using.

We are going to use our decision tree model from an earlier chapter. Below is the code for developing the model

```
library(ISLR);library(caret);library(rpart);library(rpart.plot)
data("Wage")
inTrain<-createDataPartition(y=Wage$jobclass,p=0.7, list=FALSE)
trainingset <- Wage[inTrain, ]
testingset <- Wage[-inTrain, ]
```

We are now going to develop the model for the decision tree.

```
set.seed(1)
TreeModel<-train(jobclass~health_ins+age+wage+education+maritl+race+health,
method='rpart', data=trainingset)
TreeModel
## CART
##
## 2101 samples
##    7 predictors
##    2 classes: '1. Industrial', '2. Information'
##
## No pre-processing
## Resampling: Bootstrapped (25 reps)
## Summary of sample sizes: 2101, 2101, 2101, 2101, 2101, 2101, ...
## Resampling results across tuning parameters:
##
##   cp          Accuracy   Kappa
##   0.02941176  0.6163839  0.22843967
##   0.06666667  0.5987062  0.18505281
##   0.15588235  0.5502573  0.08401039
##
## Accuracy was used to select the optimal model using  the largest value.
## The final value used for the model was cp = 0.02941176.
```

When we see the output for the "TreeModel" you can clearly see that several models were made. The "cp" hyperparameter was set at three different values by the "caret" package.

The best model is saved as part of the "TreeModel" object. When you make predictions, the "caret" package automatically calls the "finalmodel" and uses it for the test set. Below is the code.

```
testmodel<-predict(TreeModel, newdata = testingset)
confusionMatrix(testmodel,testingset$jobclass)
## Confusion Matrix and Statistics
##
```

```
##                  Reference
## Prediction      1. Industrial 2. Information
##   1. Industrial          366            207
##   2. Information          97            229
## 
##                Accuracy : 0.6618
##                  95% CI : (0.6299, 0.6928)
##     No Information Rate : 0.515
##     P-Value [Acc > NIR] : < 2.2e-16
## 
##                   Kappa : 0.3181
##  Mcnemar's Test P-Value : 4.063e-10
## 
##             Sensitivity : 0.7905
##             Specificity : 0.5252
##          Pos Pred Value : 0.6387
##          Neg Pred Value : 0.7025
##              Prevalence : 0.5150
##          Detection Rate : 0.4071
##    Detection Prevalence : 0.6374
##       Balanced Accuracy : 0.6579
## 
##        'Positive' Class : 1. Industrial
## 
```

Custom Tuned Model

Now we will try to customize the tuning ourselves to see if we can develop a better model.

For our custom tuning, we are going to add another option called cross-validation. Cross-validation involves partitioning of the data into several "folds." The model is then trained on each fold and the results are usually average from k-folds. For our purposes, we will have ten folds in the dataset.

In addition, we need to tell R how to pick the best model. There are several choices but we will select the "oneSE" which stands for within one standard error of the best performing model. All this needs to be stored in an object using the "trainControl" function.

```
ctrl<-trainControl(method='cv',number = 10,selectionFunction = "oneSE")
```

We now need to create a grid in which we store manually the different values for our hyperparameter "cp" that we want to test.

```
grid<-expand.grid(cp=c(.001,.003,.01,.03,.07,.1,.15))
```

Now we combine these two objects by adding them as arguments inside the "train" function and train our model as before.

```
set.seed(1)
TreeModelcustom<-train(jobclass~health_ins+age+wage+education+maritl+race+health,
method='rpart', data=trainingset,trControl=ctrl,tuneGrid=grid)
TreeModelcustom
```

```
## CART
## 
## 2101 samples
##    7 predictors
##    2 classes: '1. Industrial', '2. Information'
## 
## No pre-processing
## Resampling: Cross-Validated (10 fold)
## Summary of sample sizes: 1891, 1891, 1890, 1891, 1891, 1891, ...
## Resampling results across tuning parameters:
## 
##   cp     Accuracy   Kappa
##   0.001  0.6064004  0.2114791
##   0.003  0.6040329  0.2045125
##   0.010  0.6378470  0.2727983
##   0.030  0.6202257  0.2366731
##   0.070  0.5878334  0.1608480
##   0.100  0.5902031  0.1629540
##   0.150  0.5902031  0.1629540
## 
## Accuracy was used to select the optimal model using  the one SE rule.
## The final value used for the model was cp = 0.01.
```

The printout clearly shows our resampling (cross-validated 10 fold) as well as our cp values.

Now for the prediction aspect

```
testmodelcustom<-predict(TreeModelcustom, newdata = testingset)
confusionMatrix(testmodelcustom,testingset$jobclass)
## Confusion Matrix and Statistics
## 
## 
##                  Reference
## Prediction        1. Industrial 2. Information
##    1. Industrial            349            176
##    2. Information           114            260
## 
##                Accuracy : 0.6774
##                  95% CI : (0.6458, 0.7079)
##     No Information Rate : 0.515
##     P-Value [Acc > NIR] : < 2.2e-16
## 
##                   Kappa : 0.3516
##  Mcnemar's Test P-Value : 0.0003409
## 
##             Sensitivity : 0.7538
##             Specificity : 0.5963
##          Pos Pred Value : 0.6648
##          Neg Pred Value : 0.6952
##              Prevalence : 0.5150
##          Detection Rate : 0.3882
##    Detection Prevalence : 0.5840
```

```
##         Balanced Accuracy : 0.6751
## 
##          'Positive' Class : 1. Industrial
## 
```

A small 2% improve in accuracy. However, for a data science competitions this can be a major improvement in your ranking

Conclusion

Model tuning is the detail-oriented end of task aspect of machine learning. Here is where you try to make a good model a great one. The caution is always to avid overfitting by getting perfect results on the training set. This could lead to problems when using the model with new data. Therefore, there should be some restraint in the tuning of a model.

Machine learning has become one of the hottest and fastest growing disciplines of the 21st century. This has resulted in a huge demand for understanding and the use of machine learning ideas and algorithms. In this text, Darrin Thomas provides explanation and examples of the implementation of machine learning algorithms using R. Various concepts such as, classification, numeric prediction, model evaluation, and model performance are discussed.

Darrin Thomas, PhD, is Lecturer at Asia-Pacific International University located in Thailand

ISBN-13: **978-1976556814**
ISBN-10: **1976556813**

SuJinSoLa

www.ingramcontent.com/pod-product-compliance
Lightning Source LLC
Chambersburg PA
CBHW082345220526
45470CB00008B/2643